DATE			

SPACE
TECHNOLOGY
&
PLANETARY
ASTRONOMY

SCIENCE, TECHNOLOGY, AND SOCIETY

Ronald N. Giere and Thomas F. Gieryn, General Editors

JOSEPH N. TATAREWICZ

SPACE TECHNOLOGY & PLANETARY ASTRONOMY

Indiana University Press

BLOOMINGTON & INDIANAPOLIS

 ™

Manufactured in the United States of America

Library of Congress Cataloging-in-Publication Data

Tatarewicz, Joseph N.
Space technology and planetary astronomy / Joseph N.
Tatarewicz.
p.   cm. — (Science, technology, and society)
Includes bibliographical references.
ISBN 0-253-35655-5 (alk. paper)
1. Planetology—United States.   2. Planets—Exploration.
3. Astronomy and state—United States.   I. Title.   II. Series:
Science, technology, and society (Bloomington, Ind.)
QB602.9.T37   1990
523.4—dc20   89-45469
                                                           CIP

1 2 3 4 5   94 93 92 91 90

CONTENTS

Acknowledgments

In researching, documenting, and writing this study I have incurred many debts, only a few of which can I specifically acknowledge here. I hope that the many friends and colleagues who have helped me along the way already know how grateful I am to them, even if they are not mentioned here by name.

My parents Joseph and Florence Tatarewicz in Baltimore were and continue to be ever encouraging throughout my educational career, and supported this research in many ways.

Through the faculty and my fellow graduate students at Indiana University I received a fine and thorough grounding in traditional history of science and much support and forebearance as I struck out into very recent history. My advisor in the History and Philosophy of Science Department, Victor Thoren, was ever patient and helpful throughout. Linda Wessels and Ronald Giere of that department gave me valuable advice. From Ed Lambeth of the Indiana University School of Journalism I learned many aspects of science journalism which proved valuable in researching so recent a history. Martin Burkhead and Frank Edmondson of the Indiana University Astronomy Department encouraged and helped me in the scientific as well as historical aspects of this study. Thomas F. Gieryn of the Sociology Department has been a valued colleague and friend for many years.

Research grants from the Indiana Academy of Sciences and the Indiana University Graduate School allowed me to travel to Washington, D.C. for access to archival sources in the early stages of this research. In the summer of 1980 I was fortunate to be an intern at the NASA History Office, where Monte Wright, Lee Saegesser, Alex Roland, and other staff members introduced me to researching in federal archives.

Since 1982 my professional home has been the Smithsonian Institution's National Air and Space Museum. There a succession of fellowships and later a series of duties and professional posts alternately sustained and interrupted work on this research. I am grateful to the current administration for allowing me sabbatical time to finish work on this study.

My colleagues and friends at that institution, especially in the Department of Space History, helped and encouraged me more than I can express. Allan Needell and Martin Collins taught me much about organizing and using raw documentation, and about being attuned always to the highest historical standards and values. David DeVorkin challenged me to prove my hypotheses, and directed me to caches of documentation. He introduced me to the techniques of oral history, and supported the processing of interviews conducted through the Space Astronomy Oral History Project. Visiting scholars, including John Logsdon, Walter McDougall, Gerald Wasserburg, Leo Goldberg, Martin Harwit, Ron Döel, and others helped me find and interpret documentation and valuable secondary sources, including their own unpublished material. A very special thanks is due to Deborah Warner at the National Museum of American History, who took an early interest in my research, and provided valuable advice and encouragement. Karl Hufbauer and John Lankford read the entire manuscript at an early stage, and provided much valuable advice and encouragement.

Archivists and records managers at several institutions helped me locate and use a great deal of source material. I especially thank Jean St. Clair and David Saumweber of the National Academy of Sciences Archives for guiding me through the voluminous and exquisitely organized Space Science Board materials, and J. Merton England, National Science Foundation Historian, for helping me locate crucial documents.

Susan Cozzens and the National Science Foundation Division of Policy Research and Analysis supported some of the statistical work which appears in chapter five. My friend and collaborator in that effort, Tom Gieryn, worked long and hard to bring his rich and valuable expertise and his painstakingly assembled data base to bear on interesting social and institutional issues. William Greene, Charles Carter, and Joseph Berkan at NASA provided much assistance in assembling additional data, as did Lee Brakeiron at NSF.

Many scientists, engineers, and NASA managers took time out from very busy schedules to suffer my incessant questioning and requests to see documentation still in their file cabinets. Among the many, especially generous were William Brunk at NASA headquarters and Ray Newburn at the Jet Propulsion Laboratory.

An especially fond and most important thanks must go to a dear friend and companion, Janice Chase, who for many years endured and encouraged my work on this project.

Yet it goes without saying that I take full responsibility for the use to which I have put all this fine help. Any errors of fact or interpretation are entirely my own.

Introduction

"WHERE ARE THE PEOPLE WHO KNOW WHAT THEY ARE DOING?"

In March of 1959 the Sixth Lunar and Planetary Exploration Colloquium convened in California. An outgrowth of lunar base studies being conducted by the Missile Division of North American Aviation, the ad hoc series of colloquia brought together aerospace engineers and scientists to discuss the state of lunar and planetary research. Albert R. Hibbs, soon to become chief of the new Space Science Division at NASA's Jet Propulsion Laboratory (JPL) told the group of the difficulty he was having recruiting lunar and planetary expertise for a planned series of solar system probes. Hibbs lamented, "There's money available; there's interest. . . . Where are the people who know what they are doing?[1] The obvious answer to Hibbs's question would appear to be to ask the astronomers. But, NASA's L. D. Jaffe explained at a later colloquium,

At the JPL, we have been trying to get astronomical groups and individual astronomers to direct some of their efforts to the study of the planets. But they have a great deal of scientific integrity. They are interested in the galaxies and won't be diverted by any financial inducements. So putting out money is not the sole answer to the problem. We are short of astronomers in this country, and their interest, perhaps rightly so, is other than planetary astronomy.[2]

In order to design lunar and planetary probes, calculate trajectories, and develop scientific instruments, NASA planners had hoped to enlist the expertise of a science that had spent literally thousands of years preoccupied with the planets. Yet on the very eve of planetary exploration astronomers showed little interest. Moreover, this lack of interest continued even as exploration of the moon, Venus, and Mars began.

The bewilderment of NASA planners, who as nonastronomers were for the most part unfamiliar with the state of astronomy and its chief contemporary concerns and active research areas, could perhaps be excused. In the wake of Sputnik all areas of space research and exploration were confused with being "astronomical," since, as Gerard de Vaucouleurs put it, astronomers had "been in the space business for 2,000 years."[3]

Yet NASA, assisted by a diverse collection of planetary enthusiasts, pressed on to explore the solar system, and radically transformed the study of solar system bodies. Direct access to the planets by space probes allowed observations impossible from earth, setting the study of the solar system on a new observational and theoretical footing. It seems that every account of the discoveries of planetary probes begins by recounting how miserable was our knowledge of the subject before space techniques. This perspective has merit, but there is another side less often discussed. Study of the planets did not supplant ground-based "classical" planetary astronomy. On the contrary, the exploration of the solar system was prosecuted on the basis of ground-based study, and provided the means by which astronomers and others rediscovered the merits—and limits—of "old-fashioned" ground-based planetary astronomy.

My primary aim in this study is to show how space technology in a nationalistic and political context helped reinvigorate a specialty within astronomy, a specialty that many astronomers contend had fallen into disrepute. When NASA planners decided to engage in a program of exploration of the solar system from space probes, they were unaware of the neglected state of solar system studies. When finally convinced of the value of ground-based planetary astronomy, and of the improbability of the needed research being done without direct NASA support, the agency responded with a multifaceted program that transformed the field. NASA constructed observatories, provided telescopes and instruments, trained astronomers, funded research programs, and supported a host of other activities, all aimed at increasing the fund of ground-based knowledge of the planets. These developments had profound consequences for astronomy in general as well as for planetary studies in particular.

In so doing, a national government conjured into existence an entire scientific and technological system of sophisticated instruments and highly trained and motivated people to use them. An obscure backwater of astronomy suddenly became a potential tool for state ends, in an example of what Walter McDougall has elaborated as space age technocracy.[4] But at the same time state programs suddenly became a potential tool for planetary enthusiasts to pursue their own ends.

This study suggests that NASA did its job all too well. It succeeded in finding a few, and creating many more, "people who know what they are doing" in planetary research. This it did with their willing cooperation and with the expectation that planetary exploration and planetary science would continue to flourish. With proper tools and active encouragement, this young community of irreverent and promiscuously interdisciplinary scientists experienced the euphoria of exploring new worlds as a normal state of affairs, and came to think of it as their birthright.

Hence, it was with indignation, anxiety, and outright anger that they and the press confronted NASA officials, including administrator James Beggs, in the summer and fall of 1981. Beggs came to the Jet Propulsion Laboratory in August to celebrate Voyager 2's encounter with Saturn, and found himself in "the most adversary, downright hostile press conference anyone had ever seen at JPL."[5] At the Pittsburgh meeting of the American Astronomical Society's Division for Planetary Sciences in October this community confronted NASA officials and agonized among themselves. In the years since the Viking landing on Mars in 1976, no new planetary exploration program had been started. Space science budgets, especially the ground-and space-based components of the planetary program, had been declining precipitously. At the height of triumph, while this remarkable spacecraft sent back observations stunning enough to make normally sober scientists positively giddy, the scientists who had staked their professional lives on a single patron confronted their own extinction. Among the draconian measures being discussed at the time by the new Reagan administration were to simply shut down still-operating planetary spacecraft, and perhaps even get NASA completely out of the space science business.

Most of the NASA officials and the planetary scientists themselves were but dimly, if at all, aware of the implicit social contract their predecessors had made more than two decades earlier. Succeeding chapters will show the immense resistance the proponents encountered in building the base of knowledge and expertise required to explore the solar system, and the difficulty with which they overcame this resistance through a variety of tactics. Hibbs and others had stood before groups of astronomers and implored them to enter a field of study for which there were few inducements. Now Beggs and others stood before those who had answered the call and staked their professional lives on the stability of the NASA program. Hence this study concentrates on, but is not limited to, the years between Hibbs's call in 1959 to would-be explorers of the solar system and their successors' confrontation with NASA officials in 1981.

Perhaps in another age, and perhaps in other sciences and exploratory endeavors, the researcher chose an area in which to work and took his or her chances. But the community of planetary scientists who confronted the NASA officials came of age when government support and management of science and technology were for them a given, and they felt abandoned. The events recounted in this study thus raise interesting questions of ethics and responsibility in state-managed research and development. Each of the proponents who encouraged scientists and students to turn their attention to planetary study and encouraged universities to found entire departments and programs did so with the sincere belief that such activity would have a secure future. Was it reckless to lure people into careers that were viable only insofar as the state continued to send spacecraft into the solar system? Or were the explorers naive to believe they could sustain their expensive and elaborate voyages on the basis of scientific worth alone? David Morrison, one of the most politically adept and historically aware of planetary researchers, must have been engaging in delicate rhetorical exhortation when he said in a letter to his colleagues in 1981, "The time has come to politicize the planetary science community. . . . I realize this is distasteful to many, and I am sure all of us would much rather pursue our science."[6] As succeeding pages will show, their community was thoroughly politicized from the beginning. And as much recent literature in the social studies of science and technology contends, the lines between science, technology, politics, and society are anything but fixed and impermeable.

PLAN OF THE BOOK

The immediate stimulus for increased planetary astronomy came from the lunar and planetary flight programs. But just as the expanded space program after 1957 capitalized on scientific, technical, political, and social forces which had been gathering since World War II, so too the renewed activity in planetary astronomy had roots stretching back at least as far. NASA drew on the expertise of individuals and institutions who had long been active in planetary research. The state of planetary astronomy, purported reasons for its neglect during the early twentieth century, and the increasing interest in this specialty shown by various parties outside of astronomy between World War II and the creation of NASA in 1958 form the subject of chapter 1.

Once remote objects of scientific curiosity, the planets took on new and more immediate significance in the near hysteria that followed in the wake of the Soviet launch of the Sputnik satellite. Segments of NASA saw in the planets an opportunity to recover lost prestige; parts of the military saw the moon and even the nearer planets as the next arena of battle; aerospace companies saw opportunities for lucrative contracts; engineers working in various aspects of rocketry saw interplanetary travel as an attractive challenge to their skills. These new motivations for studying the planets and the expanding demand for planetary astronomers are discussed in chapter 2, as well as the more traditional scientific interest in understanding the solar system and the earth's place within it. Attempts to convince NASA and its advisory bodies that an ongoing program of planetary astronomy was required to support the planetary exploration program are described.

Once the appropriate officials had been convinced that NASA needed to support the ground-based study of the planets in order to successfully execute its various probe missions, questions of the legitimacy of such support still needed to be settled. Certain officials questioned the need for the space agency to support ground-based studies, which were the province of the National Science Foundation (NSF). Many scientists feared the intrusion of the agency into the orderly setting of scientific priorities, and questioned the scientific motivations of NASA's program. The severe demand that the space program was placing on astronomy in the mid-1960s, the shortage of astronomers relative to the demand, and the crowded and inadequate observing and research facilities all helped make astronomers more receptive to NASA's involvement in providing ground-based facilities. These issues, and the events which figured in their resolution, are discussed in chapter 3.

The overriding emphasis placed on Project Apollo diverted ambitious plans for planetary exploration made in the late 1950s. Around 1965, as planners anticipated the peak and decline of Apollo funding, internal and external planning bodies converged on an ambitious program of solar system exploration as a post-Apollo goal. NASA expanded its support of ground-based planetary astronomy during this period, constructing optical telescopes and training astronomers to undertake planetary studies. This consolidation of planetary astronomy as a secure and ongoing part of the planetary exploration program forms the subject of chapter 4.

The optical facilities which NASA had supported in the first five to eight years of the planetary astronomy program helped alleviate the shortage of telescope time which planetary astronomers had faced earlier. The agency's emphasis shifted in the 1970s toward construction and renovation of radio and radar facilities for planetary astronomy, and the construction of a major infrared telescope. The research emphasis of the program also shifted. Whereas Venus and Mars had been emphasized in the earlier period, the new targets for the planetary exploration program—Mercury and the outer planets—required increasing ground-based reconnaissance. These new emphases in facilities construction and research support are also treated in chapter 4.

The growing community of planetary researchers, their multidisciplinary allegiances, and the peculiar combination of space- and ground-based research techniques brought to bear on the planets called for novel institutional arrangements. The attempts to find a "home" for planetary science within existing scientific societies and associations are described in chapter 5, along with the allied problem of finding or creating appropriate publishing outlets for research results. The results of this study of planetary astronomy are compared to similar studies of x-ray astronomy, radio astronomy, and radar astronomy, and other views of NASA's role in the development of space sciences are examined.

CONCLUSION

This is about one of many possible histories of planetary studies during the period between 1959 and 1981. It is primarily the story of a small scientific specialty suddenly elevated to prominence and importance by events outside of science, and the response of interested parties to this new importance and affluence. It is not meant to be a detailed history of planetary science or even of planetary astronomy during the period in question. Although such a history would be valuable and interesting, it must await the completion of many historical studies of smaller scope. With the present study, I hope to contribute one such piece.

I concentrate on the NASA program of research support and instrument development which sought to alleviate the perceived shortage of planetary astronomers, facilities, knowledge, and interest in support of a broader program of solar system exploration. I attempt to set this program within its proper historical context, showing how it arose, how it interacted with the many other elements of science, technology, politics, and economics with which it was interlinked, and the effect it had on the disciplinary economy of science in the space age.

Nor is this study meant to buttress one of the several interpretive frameworks of current sociology of science and technology, political science, or science policy. While at various occasions in the narrative, and more explicitly in the concluding chapter, I find that one or another interpretation helps illuminate my account, no single such framework can fully explain the complexity of these many events and personalities. The story I recount happened but once, under very specific and contingent historical circumstances. By making this story available to a wider critical audience through what is primarily a traditional historical narrative, I offer only a better understanding of what took place, and material for whatever lessons for policy or theory the reader cares to make it serve.

Space
Technology
&
Planetary
Astronomy

Planetary Astronomy in the Age of Astrophysics (1900–1958)

STATE AND STATUS OF PLANETARY ASTRONOMY

A large body of contemporary planetary astronomy literature exists in the form of conference proceedings and compendia. Often, as an introduction or a preface, a brief sketch of the history of planetary astronomy in the twentieth century is offered, from the viewpoint of the practicing planetary astronomer. While such accounts serve many purposes, and the history contained therein is shaped by the purposes of the author, insofar as they express the lore of the practitioners it is worth examining some representative examples. For it was this version of history that the proponents of planetary astronomy used to buttress their cases for support from NASA and other institutions, and to indoctrinate new recruits.[1]

Almost without fail, such resumés agree on several points. First, study of the moon and planets, once the major part of astronomy, was almost entirely abandoned in the early part of the twentieth century. Second, the new opportunities that opened in astrophysics and galactic astronomy during this period absorbed most of the interest, energy, and resources of professional astronomers. Third, the controversies over Mars and Percival Lowell's interpretations helped push visual planetary astronomy into disrepute among professionals, after which such study of the planets became almost entirely the province of amateur astronomers. Fourth, the limitations of available instrumentation and theory made study of planetary physics much more difficult than other astronomical topics. Finally, it was the space program that revived ground-based planetary astronomy and revolutionized the overall study of the planets by introducing new observational techniques and theoretical interpretations from other sciences.

Gerard P. Kuiper, a distinguished stellar astronomer who later increasingly devoted his energies to solar system studies, attributed the "almost complete abandonment of planetary studies" early in this century to the factors mentioned above. But even though stellar and galactic astronomy held center stage, a few astronomers "who occasionally took time off for planetary studies (at the risk of mild scorn from their colleagues) found a rich field for investigation," as they applied to the planets techniques developed primarily with stellar studies in mind.[2]

Walter Fricke recalled that when he was a graduate student in Berlin and Göttingen during the late 1930s, "the senior astronomers told their students that the planetary system does no longer offer a subject for astronomical research which would lead to significant new findings. The planetary system was considered to be well known such that further exploration could at best lead to minor refinements in the description of some well-established facts."[3]

Neville J. Woolf identified planetary astronomy as a field which before the space age was "perceptibly yawning," and has "sunk into a deep lethargy." After discussing the technical difficulties of planetary astronomy, Woolf noted that "these difficulties caused planetary astronomy to fall into disrepute. The observations that were made did not convert into facts. . . . As a result, speculation flourished unchecked by observation. The widespread public interest has also caused difficulties, since it has tended to encourage cranks to enter this field which is difficult enough without them."[4]

These examples show a widespread belief among practitioners of planetary science that their field suffered a severe erosion of prestige and support around the time of the great controversies over Mars, a belief that is confirmed by some historical studies.[5] In all of these issues, the controversial figure of Percival Lowell and his observatory in Flagstaff, Arizona loom large.

THE LEGACY OF PERCIVAL LOWELL

The first half of the twentieth century was a period of intense change for astronomy. The field was in transition from being primarily an application of gravitational theory to planetary motions, to the beginnings of a true physics of heavenly bodies. The new physical methods (as well as the theories of physics which would help interpret the results) were still controversial and evolving. This prompted a certain amount of uneasiness and debate about the proper subjects and methods of astronomy. Agnes M. Clerke, writing in 1902, described the displeasure of the "orthodox astronomers of the old school [who] looked with a certain contempt upon observers who spent their nights in scrutinizing the faces of the moon and planets rather than in timing their transits. . . ."[6] The histories which treat twentieth-century astronomy, and especially astronomical textbooks, give center stage to the success stories of the new astrophysicists who eventually applied the new physics to the stars.[7] Yet at the same time that astrophysics and galactic astronomy were displacing other fields in importance and emphasis, an enthusiastic group of astronomers in Arizona made physical study of the planets their primary occupation.

In the 1890s spectroscopy was in its infancy, astronomers were slowly warming up to the use of photography, physical methods were beginning to displace purely positional astronomy, and astronomers were just beginning to professionalize. Into this atmosphere came Percival Lowell, a wealthy and mathematically educated businessman who, upon hearing of Giovanni Schiaparelli's failing eyesight, resolved to continue the Italian's observations of Mars. Lowell had enthusiasm, money, intense energy, a gift for clear and imaginative writing and speaking, and astronomical ability and acquired knowledge perhaps equal to any professional astronomer. He had resolved to devote himself to the study of Mars and the other planets during the favorable opposition of Mars in 1892, some years after Schiaparelli's discovery of the "canals," when

> a veritable tumult began. Novels about Mars poured from the pens of imaginative writers. Telescopes were built for the specific purpose of observing [this] opposition. Mars observatories sprang up by the dozens. Amateurs detected light signals. Millionaires offered prizes for communicating with the Martians. It was proposed that a drawing of the Pythagorean theorem half the size of Europe be traced on the Sahara Desert that Martian astronomers might see it and realize that the creatures of Earth were also intelligent. Never had there been such a wild enthusiasm for astronomy.[8]

Lowell established his observatory in Flagstaff, equipped it with the finest instruments available, recruited a string of remarkably able staff members, several of whom remained intensely loyal through the years, and attacked Mars and the other planets with every technique at his disposal.[9]

Although Lowell's astronomical work was in many ways innovative, and he pioneered many techniques which were to become a mainstay of planetary astronomy even through the NASA years, the controversy over interpretation of the observations was at times bitter. R. L. Waterfield, writing in 1938, characterized the story of the "canals" as a "long and sad one, fraught with backbitings and slanders; and many would have preferred that the whole theory of them had never been invented."[10] Many of Lowell's observations of Mars and other planets were not confirmed at other observatories with larger telescopes. In reply, Lowell questioned the quality of their instruments, the environmental conditions at the observatories, the methods used, and even the competence of some of the most powerful and respected astronomers of the day.[11] He several times claimed to have photographed the "canals," but could not present the evidence because of the inadequacy of printing techniques of the day; he exhibited the photographs, but few could see the hints of linear features.[12] In any case, no photograph could catch the detail which a trained observer could see during instants when the atmosphere calmed, so it was often a case of believing the observer. The fascination of the public with Mars, which perhaps reached its peak at the favorable 1909 opposition, complicated matters. Reports of "discoveries" were garbled by the press, and other astronomers sometimes first learned of Lowell's various claims through such misleading reports.[13]

There are two views on the effect of all of this, both on the Lowell

Observatory staff and on planetary studies in general. Waterfield stresses Lowell's positive contribution to planetary studies, claiming that whatever harm may have been done by the controversy was outweighed by the stimulus that it gave to physical studies of Mars and other planets. "Whether in a positive way to champion it, or in a negative way to oppose it, it attracted many able observers who otherwise might never have taken an interest in the planets. . . . So the pistol which Schiaparelli so unwittingly let off, though it shocked the finer feelings of many, had undoubtedly been the starting signal of that race for discovery which the planetary astronomers are still successfully pursuing."[14] Otto Struve and Velta Zebergs, writing in 1961, concurred, calling Waterfield's summation "an excellent final comment on the Mars controversy." They gave credit to Lowell for the fact that, unlike the nineteenth century when planets were observed mainly by amateurs with small instruments, at the time of their writing "most large observatories devote a considerable proportion of their observing time to planetary studies and some of the best-known astrophysicists are attempting to interpret these observations."[15]

Carl Sagan and I. S. Schlovskii, however, stress the negative aspects and effect of the canals controversy, saying, "It became so bitter and seemed to many scientists so profitless, that it led to a general exodus from planetary to stellar astronomy, abetted in part by the great scientific opportunities then developing in the application of physics to stellar problems."[16]

Whatever the final judgment on this question, Lowell Observatory was the largest and best equipped institution in this period explicitly dedicated to study of the planets. In the aftermath of the Mars controversy, however, the staff at Lowell extended their research beyond the planetary system, and tempered the pace of the planetary work as well. Clyde Tombaugh, who discovered Pluto, recalled a 1929 conversation with Carl O. Lampland in which "Lampland said that Lowell urged them to engage in nonplanetary research in order to gain some credibility among stellar astronomers."[17]

After the death of its founder, the monopoly of planetary work at Lowell continued to lessen and the institution also found itself in poor financial condition. Although Lowell had left a large bequest for the maintenance of the observatory, his widow Constance contested the will, and a "complex and somewhat bitter legal battle . . . dragged on for more than ten years and not only cast a shadow over the provisions Lowell had made for his observatory in his will, but progressively sapped the resources of his estate through court costs and what the elder Slipher declared to be 'very excessive' attorney's fees."[18] Thus, although a good deal of work on Mars and other planets was done at Lowell in the years between its founder's death and the disruption caused by World War II, this "planetary astrophysics" was not prosecuted with the vigor which had been the rule during the life of its founder. Finally, the staff, especially Lampland and Vesto M. and Earl C. Slipher, engaged in a good deal of "fence mending" after Lowell's death. Tombaugh recalls that when he arrived to take up the search for a planet beyond Neptune in 1929, the long legacy of the Mars canals controversy had seriously demoralized the staff. A few times the two Sliphers and Lampland poured out to me the anguish of their souls. One time, Dr. V. M. Slipher told me, "The lot of the older men here is not a happy one," because of the ostracism from the astronomical community.[19]

William Graves Hoyt, however, gives no indication of such a gloomy climate at Lowell during those years. In fact, he found that in the years following the favorable 1924 opposition, "Lowell's much disputed postulate of a habitable Mars came as close to general acceptance as it ever would." He cites a letter from Slipher at the end of 1926 in which Slipher noted: "Years ago most astronomers looked with suspicion upon reports of observations concerning the planet and rather looked upon such work as unscientific and injurious to the reputation of those engaged in it. The Pendulum now seems to have swung oppositely." Hoyt, however, does relate that the work done in the years after Lowell's death was almost all observational, with little attempt to synthesize the data or theorize "as Lowell had done so provocatively during his controversial career." They published and reported little, and what they did make public was couched in very careful and restricted terms. "One effect of this conservatism was that the Lowell Observatory now began to acquire a reputation for accuracy and quality in its work that it had not always enjoyed when Lowell guided its affairs."[20]

Thus it seems clear that Lowell's legacy to planetary astronomy is complex. He and his assistants pioneered many astrophysical techniques such as polarimetry, spectroscopy, radiometry, and photography in the study of the planets. NASA planners and planetary astronomers through the 1960s and 1970s would return again and again to the results and techniques of these early studies, undertaken by or in response to Lowell. The enormous collection of photographic plates and hand-drawings of all of the planets would become most valuable in later years to researchers looking for longterm changes and patterns in meterological phenomena. It is also clear, however, that planetary astronomy had developed a reputation of being observationally difficult, and uncertain in its tendency to theorize from scanty observational evidence. The hazards of such premature theorizing, however, have to be learned again and again into the space age.

If the ultimate goal of planetary astronomy is the complete "geophysical" characterization of the planetary system, then planetary astronomy in the first half of the twentieth century advanced slowly when compared with the dramatic achievements of stellar and galactic astronomy and cosmology. Indeed, more was known at the time about the physical constitution and evolution of the stars than of planetary bodies. In part this was due to the inherent difficulty of the observations, the ambiguity of the results and their overinterpretation, the rather poor understanding of the earth as a planet, and the relatively few researchers in the field. In part it was due to the rather unpleasant experience of the Mars controversy, which made researchers wary of entering such a line of research. The chief modern techniques of planetary study, however, *were* exercised and their limitations probed. World War II would substantially disrupt most astronomical research, planetary as well as stellar and galactic. But as the war came to a close, the tools and institutions which would transform all of astronomy were beginning to appear. Radio and radar astronomy, alternatives to the photographic plate, spaceflight technology, and increased federal involvement in science would soon combine under the catalysis of international political developments to draw far greater attention to the moon and planets.

PLANETS RISING

Interest in the atmospheres, surfaces, and interiors of the planets and other solar system objects, once almost exclusively confined to astronomers, arose among a number of other parties in the postwar period. Some of these groups came to be interested in the planets as a result of the progress of their own disciplines. Other groups developed interest in the planets for nonscientific reasons. Astronomers too showed some increasing interest in planetary studies, partially as a result of developments in cosmogony and cosmology, partially as a result of new instrumentation available as a by-product of war research. In many cases, the increasing interest in the planets was in some way associated with the realization that rocket propulsion was on the verge of practicality.

ASTRONOMERS DURING THE WAR

Struve and Zebergs began their *Astronomy of the Twentieth Century* with a chapter called "The Fortune of Astronomers," in which they noted that before World War II astronomy "was the purest of sciences," poorly funded, and small in size and activity. After the war, however, it "emerged as a science of national importance."[21] The negative impact of the war on the activity in astronomy is illustrated by Struve and Zebergs' graph, which shows the number of active astronomers dipping from 2,000 to less than 1,000 during the war years and then exceeding 3,000 by 1955.

During the war astronomers served in a number of capacities whereby their unique capabilities could be used: "From ballistics, navigation, and optics to nuclear physics, rocketry, ship magnetism, undersea reconnaissance, cryptography, electronics, and project management, astronomers acted as applied physicists, logicians, and mathematicians."[22] While the war disrupted astronomical research, it brought many benefits to the field afterward.

Bernard Lovell has written of the two decades after the war as an "epoch when defense science not only influenced but actually revolutionized observational astronomy." He identified three critical events: the development of radar, the growth of scientific personnel after the war, and the launching of Sputnik. Lovell went on to specify how scientists' wartime experience in various defense research units had affected them, writing, "At the end of the war they were themselves acquainted with the most sophisticated and advanced techniques of the day. They had learnt how to collaborate with a wide range of scientific and non-scientific disciplines, and they had learnt the way of committees and how to use government machinery."[23] The new relationships forged between government, military, and science during the war altered the conduct of science. The military, for its part, discovered the value of scientists and emerged from the war with a desire to maintain the productive alliance. The scientists emerged from the war as Lovell describes them, but also with a desire to return to their interrupted research, to regain the time lost, and to return to the university atmosphere. They also emerged with a taste for large budgets.[24]

Planetary astronomy benefited from the war in a number of ways. The increased availability of money helped alleviate the financial pressures that had constrained all of astronomy, and made it somewhat easier to pursue marginal studies, such as planetary astronomy was at the time. New instrumentation developed during the war helped open up the infrared wavelengths beyond the photographic range, a spectral region important for discerning molecular constituents of planetary atmospheres. Radar techniques opened up the microwave and shorter wavelength regions, allowing radiometry to be applied with far more precision and flexibility than was possible before the war. Wartime experience had drawn increasing attention to various scientific problems on the boundaries of astronomy and geophysics: the structure, composition, and optical properties of the atmosphere; the nature and dynamics of the aurora and skyglow and their connection with solar activity; the dynamics of the earth's weather and global circulation patterns; and many others. Deeper understanding of the earth as a planet would help frame questions to guide research into the characteristics of other planets, while the vastly different conditions on those planets would invite investigations which in turn would deepen understanding of the earth. Finally, rocket and missile technology, and the political nexus in which it was embedded, would in time make travel to the planets a feasible and attractive goal for a variety of reasons.

MILITARY INTEREST

In 1949 the newly independent U.S. Air Force assumed funding of a project at Lowell Observatory to study planetary atmospheres. Initially an undertaking of the U.S. Weather Bureau in 1948, the project was aimed at clarifying the large-scale circulations of the earth's atmosphere through the study of the atmospheres of Mars and Jupiter. The researchers hoped to overcome a severe problem in presatellite meteorology: distinguishing fluctuations in observations caused by large-scale phenomena from those of a purely local nature. By observing the global circulation patterns of other planets, and using the existing plate files of the Lowell Observatory's fifty years of extensive planetary observation, the researchers hoped to use the planetary perspectives as "miniatures of the standard daily weather maps of the earth."[25] The project brought together meteorologists and astronomers in a collaborative effort to review existing observations and conduct many new observations. Two major conferences were held, one on planetary atmospheres in March 1950, and the other on solar variation and planetary atmospheres in August 1951. Attending scientists and members of the project are listed in tables 1.1–1.3.

The planetary atmospheres project made good use of the enormous quantity of undigested basic data that had been amassed at the Lowell Observatory, and helped keep the institution involved in planetary research. The war years had been particularly hard for Lowell, and although the planetary work was kept up by Lampland and the Sliphers, most of the other staff such as Tombaugh, James Edson, and Henry Giclas had been drafted into military service. The observatory's financial difficulties seemed chronic, and continued into the 1950s. Only Giclas returned after the war, and he worked increasingly on stellar photometry. The planetary atmospheres project reinvigorated the observatory's planetary research, introducing several meteorologists to the potential of ground-based planetary observations for understanding the earth's weather. Yale Mintz and Seymour Hess especially would in later years take all the planets with atmospheres as their domain of study, and contribute in significant ways to NASA's program of planetary exploration and ground-based supporting research.[26]

The military in general, and the Air Force especially, had good reason to be interested in all aspects of weather prediction. This interest had a long history, although the notion that observations of the weather on other planets might help clarify concepts of the earth's meteorology was new.[27] Wartime experience had also shown the significance of an understanding of the dynamics of the earth's atmosphere for communications and radar, and had suggested ties between those dynamics and variations in solar radiation. The Air Force established the Sacramento Peak Solar Observatory at Sunspot, New Mexico for the purpose of monitoring and studying solar activity and its connection to events in the earth's atmosphere and ionosphere.[28]

In addition to communications and the effect of weather on military operations, a new motivation for increased attention to the characteristics of the earth's atmosphere arose after the war. The Air Force considered space itself a logical extension of its arena of operations, and was keenly interested in conditions in the upper atmosphere and the space beyond which could affect communications, guidance, and control of air and space vehicles, as well as the various effects of such an environment on pilots and passengers.[29]

The development of lightweight nuclear warheads in the early 1950s made ballistic missiles an attractive delivery system, and development work on such rockets was well underway by the Air Force and Army as well. Guidance and control calculations started with basic celestial mechanics, and thus geodetic and geophysical description of the earth's body was required to correct the ideal trajectory for perturbations introduced by variations in the mass distribution of the earth. The effect of the resistance

TABLE I.I

Scientific Staff of the Lowell Observatory–Air Force Planetary Atmospheres Project, 1948–1952. (Duration of work with the project varied. See administrative summary in final report for details.)

Earl C. Slipher, Lowell Observatory
Astronomer-Supervisor
Henry L. Giclas, Lowell Observatory
Astronomer
Seymour L. Hess, Florida State University
Meteorologist
Ralph Shapiro
Meteorologist
Edward N. Lorenz, MIT
Meteorologist
Alfred K. Blackadar
Meteorologist-Consultant
Franklin A. Gifford
Meteorologist
Yale Mintz
Meteorologist

TABLE I.2

Conferees at Lowell Observatory Conference on Planetary Atmospheres, March 17–20, 1950.

Arthur Adel, Arizona State College, Flagstaff
Astronomer
James B. Edson, Aberdeen Proving Ground
Astronomer
C. T. Elvey, U.S. Naval Ordnance Test Station, Inyokern
George Herbig, Lick Observatory
Astronomer
Seymour L. Hess, Lowell Observatory
Meteorologist
Gerard P. Kuiper, Yerkes and McDonald Observatories
Astronomer
C. O. Lampland, Lowell Observatory
Astronomer
Charles C. Osterberg, Lowell Observatory
Rudolph Penndorf, Cambridge Field Station, Army Materiel Command
Edison Pettit, Mt. Wilson and Palomar Observatories
Astronomer
Roger L. Putnam, Lowell Observatory
Trustee

TABLE 1.2 *(continued)*

Franklin E. Roach, U.S. Naval Ordnance Test Station, Inyokern
E. C. Slipher, Lowell Observatory
 Astronomer
V. M. Slipher, Lowell Observatory
 Astronomer

TABLE 1.3

Participants at the Lowell Observatory Conference on Solar Variation and Planetary Atmospheres, August 22–24, 1951.

Arthur Adel, Arizona State College, Flagstaff
 Astronomer
E. P. Carpenter, University of Arizona
 Astronomer
A. E. Douglass, University of Arizona
 Astronomer
John C. Duncan, University of Arizona
 Astronomer
Frank K. Edmondson, Indiana University
 Astronomer
H. L. Giclas, Lowell Observatory
 Astronomer
Seymour L. Hess, Lowell Observatory
 Meteorologist
Gerard P. Kuiper, Yerkes and McDonald Observatories
 Astronomer
C. O. Lampland, Lowell Observatory
 Astronomer
R. Long, University of Chicago
 Meteorologist
Edward N. Lorenz, University of Chicago
 Meteorologist
Roas Miller, Northrop Aircraft Corporation
 Meteorologist
Charles C. Osterberg, Lowell Observatory
Roger L. Putnam, Lowell Observatory
 Trustee
Zdenek Sekera, University of California, Los Angeles
 Meteorologist
Robert Shaldaoh, Lowell Observatory
Ralph Shapiro, Lowell Observatory
 Meteorologist
E. C. Slipher, Lowell Observatory
 Astronomer

TABLE 1.3 *(continued)*

V. M. Slipher, Lowell Observatory
    Astronomer
P. H. Taylor, Northrop Aircraft Corporation
    Meteorologist
E. Wahl, Geophysics Research Division, U.S. Air Force, Cambridge
    Meteorologist
Harry Wexler, U.S. Weather Bureau
    Meteorologist
Robert M. White, Geophysics Research Division, U.S. Air Force, Cambridge
Rupert Wildt, Yale University
    Astronomer
Jocelyn Gill, Smith College
    Astronomer

of the earth's atmosphere also had to be calculated, a resistance which changed in response to the fluctuations of the atmosphere. Finally, the guidance of the missiles was partially accomplished through radio contact, and the effects of the ionosphere on radio signal propagation had to be considered.[30]

Air Force planners did not let their hopes stop with advanced missile and air/spacecraft for flight within the vicinity of the earth. The Air Force Office of Scientific Research had by 1957 developed a program which "was frankly and overwhelmingly space oriented," and which among other things looked to "the probing of outerspace for scientific information essential to any space venture."[31]

To this end, the Office of Scientific Research sponsored annual symposia on astronautics. The first, in February 1957, included a surprisingly large proportion of papers on interplanetary travel.[32] The second symposium, four months after Sputnik, was even more frankly lunar and interplanetary in orientation. Morton Alperin, Director of Advanced Studies of the Office of Scientific Research, wrote in the preface to the proceedings that his intent was "to place in the hands of the most competent scientific people and laboratories, problems which, although of immediate importance to the Air Force, are also significant contributions to an Air Force of the future which I am certain will be required to operate in space beyond our atmosphere." Topics included the photographic reconnaissance of Mars, experiments from lunar vehicles, a panel discussion on an earth-orbital space station, and a special session on the moon.[33]

In the aftermath of Sputnik, the moon and even the nearer planets beckoned as sites for surveillance of the earth and basing of nuclear missiles. Air Force Lt. Gen. C. S. Irvine, speaking at the Western Space Conference of the Los Angeles Chamber of Commerce, called his theme, "the Air Force today is the space force of tomorrow," while California Institute of Technology President Lee Dubridge ridiculed "Sunday supplement stories that the moon would be a good observation post and launching site" and "wild programs of Buck Rogers stunts and insane pseudo-military expeditions."[34] Air Force Gen. Homer A. Boushey predicted that the moon and Mars would have permanent manned outposts within twenty years, and that aside from problems of survival during the voyage, all other aspects of a Mars base "seem to me to be easier of solution."[35] The trade journal *Aviation Week* in September 1958 reported that "top U.S. planners in the over-all defense and space technology picture are receptive to proposals for studies to tackle some of the initial problems envisioned in the long-range approach to manned lunar landings and bases. Sizable initial financial support for this work may not be too difficult to get, observers close to the picture believe."[36] In April 1959 Air Force planners heard progress reports under two separate studies then underway by industry, one for a lunar observatory for surveillance, and the other for "military bombardment capability from a moon base." In addition to various departments within the Air Force and NASA, eight aerospace firms were present, some of which were conducting the studies under Air Force contracts, others of which were conducting their own internal studies.[37]

By September the trade journal reported that thinking was going beyond the moon. Under the same study requirement, Air Force officials were exploring "strategic interplanetary systems" for "operating vehicles and weapons beyond the orbit of the moon. This implies a weapon systems capability coupled with capability for interplanetary travel. . . . One premise under which the studies are progressing is that, if bases for scientific exploration on near planets are established, an obligation would evolve to protect them. Coupled with this is the philosophy that scientific findings on planetary bases may bring out a prime necessity of holding these stations." The contractors for the lunar base studies were General Motors, Westinghouse, and Douglas Aircraft; the contractors for the interplanetary systems study were Aerojet-General, Douglas Aircraft, and North American Aviation. The journal also reported that "a number of other industry members also are performing studies in this space regime . . . in preparation for development work which is almost certain to be generated in the future by military or civilian agencies."[38]

James R. Killian, Eisenhower's science advisor immediately after Sputnik, summarized the effect of the sudden enthusiasm for space on certain segments of the military. The president had made it clear that the majority of space activities were going to be the responsibility of a civilian agency (later to become NASA). The President's Science Advisory Committee had concluded that the main military applications of space lay in surveillance and communications only. Yet,

it is strange now to recall the fantasies that *Sputnik* inspired in the minds of many able military officers. It cast a spell that caused otherwise rational commanders to become romantic about space. No Sir, they were not going to fight the next war with the weapons of the last war; the world was going to be controlled from the high ground of space. . . . And they were convinced that their service, be it army or air force, was best qualified to develop the exotic technology that would be needed for space warfare—and for civilian use too.[39]

While the studies conducted by industry never resulted in military use of the moon and planets, their serious consideration and the letting of lucrative study contracts with the promise of future work kept the military and contractors occupied with the more scientific questions concerning the characteristics of the moon and planets. The Air Force supported and conducted a wide variety of lunar and planetary research including literature surveys and publications, ground-based observatories, balloon-borne telescopic observations of the planets, and independently contracted observations and analysis by university investigators. A Lunar and Planetary Exploration Branch was established under John W. Salisbury within the Geophysics Research Directorate of the Air Force Cambridge Research Laboratories.

Three points of interpretation are important to understand, however. First, the various kinds of basic scientific research supported by the Air Force and other military services, while explicitly justified in terms of the defense mission, enjoyed a wide latitude of relevance to that mission.[40] It was a characteristic of defense-funded basic research generally that military relevance was interpreted broadly, at least during the 1950s and early 1960s.[41] Second, the military services were not monolithic in their support of space research. Individuals and groups within the defense community were to be found at all positions on the spectrum, from advocacy of "strategic interplanetary systems" to those who considered space to have minor or very circumscribed military application. The combination of relatively broad mission justification for contracted research (which allowed earth's weather to be studied by analogy with Mars and Venus) and the increasing credibility of space flight during the 1950s (which made at least lunar travel plausible) resulted in increased funding for lunar and planetary research.[42] Finally, the individual military services competed with one another for the overall defense budget, roles, and missions. All wanted to maintain competence, independent of the civilian sector, in any area of science and technology even remotely related to their missions, and each wanted to maintain such competence independent of the others.[43] In the aftermath of Sputnik, the interservice rivalries intensified, the military relevance of space seemed limited only by the imagination, and money for all manner of space research became freely available.

Similar problems and research activities occupied the U.S. Army. At the close of World War II the German rocket team, consisting of Wernher Von Braun, more than a hundred of his best engineers, and boxcars of V-2 rocket parts and documentation had been transferred from Germany to the United States, eventually arriving at the Army's Redstone Arsenal in Huntsville, Alabama. There the Von Braun team worked on the Army's ballistic missile programs. But even as Von Braun worked on the various Army missiles, he was designing and planning much larger vehicles. Interplanetary travel, and missions to Mars in particular, preoccupied Von Braun and many members of his team. Magnus Von Braun had told Army intelligence agent Charles L. Stewart that "they had selected the Americans, as they were favorably disposed toward this country generally and also because this country was the one most able to provide the resources required for interplanetary travel."[44]

In 1952 Von Braun published *The Mars Project* to "prove that we can thrust an expedition to Mars with conventional, chemical propellants." The study envisioned a flotilla of ten spacecraft, with a crew of at least seventy, assembled in earth orbit and then inserted into a trajectory taking it to Mars in 260 days, where landing craft would descend to the surface. Basically a "feasibility study," Von Braun admitted that while he had proved the technical possibility of the voyage, "true space travel cannot be attained by any back-yard inventor, no matter how ingenious he might be. It can only be achieved by the coordinated might of scientists, technicians, and organizers belonging to nearly every branch of modern science and industry. Astronomers, physicians, mathematicians, engineers, physicists, chemists, and test pilots are essential; but no less so are economists, businessmen, diplomats, and a host of others."[45]

Late in 1956 Secretary of Defense Charles Wilson issued a "roles and missions" memorandum giving the Air Force jurisdiction over the intercontinental missiles and restricting Army weapons to ranges of less than two hundred miles. The Army, in order to justify retaining the Von Braun team, had to come up with a mission for the proposed Saturn space project. Prospective uses advocated over the next three years included combat troop and cargo transport, and a manned lunar landing perhaps by July 1965. After the authorization of NASA in the spring of 1958, a "three-way tug-of-war between the Army, the Air Force, and NASA over who should manage the national large booster program, and for whom Von Braun and his team should work" lasted until Eisenhower transferred the team to NASA in October 1959.[46] As part of the campaign to keep the Von Braun team during 1959, the Army conducted a study called "Project Horizon," whose first draft in June concluded that the "earliest possible U.S. manned lunar outpost was vital to U.S. interests." A second draft, rewritten to eliminate military implications, was submitted in September. The report predicted that if the project were approved immediately, then seven years and more than two hundred Saturn launches later, an outpost occupied by twelve men could be in place on the moon. The goals of such an undertaking were: (a) to demonstrate American scientific leadership in space; (b) to serve as a communications relay station, as a laboratory for space research and development, and as a stable, low gravity launching site for deep space operations; (c) to provide an emergency staging area, rescue capability, or navigational aid for other space activity.[47] A month later President Eisenhower reversed a previous decision and transferred the Von Braun team to NASA, effectively ending the Army's bid for a large role in the space program. But the planning undertaken by the Army in its attempt to keep the Von Braun team would later prove useful to NASA in its lunar and planetary program.

The Army's interest in the space beyond the earth was perhaps more restricted than that of the Air Force. But in the course of developing ways of tracking missiles and, later, satellites, Army ordnance made a significant contribution to planetary studies. Tombaugh, who had served at Lowell Observatory from 1929 to 1943 in the search for "Planet X" and other small solar system objects, had been in charge of optically tracking the V-2 rockets fired from White Sands, New Mexico, by the Army since 1946. Tombaugh thought that the area surrounding the earth should be surveyed for possible natural earth satellites "to ascertain some of the hazards such missiles might experience, and particularly to prevent possible mis-identification with possible natural satellites." Army ordnance funded Tombaugh's survey of the space surrounding the earth to a distance of one million miles, in part to develop experience in optically tracking very high altitude rockets and satellites. Until the development of reliable radio tracking methods, optical tracking of missiles and satellites was the preferred method. If such natural earth satellites existed, they could prove a confusing nuisance to trackers. If any such natural earth satellites were discovered, tracking them would improve the geodetic description of the earth for more precise intercontinental missile targeting and aid astronomers in astrometry and celestial mechanics, and they could even serve as natural bases for a space station. The search was conducted at Lowell Observatory from 1953 to 1956 using ingenious optical techniques developed by Tombaugh during his White Sands work and planet search work at Lowell in the 1930s and early 1940s.[48] Tombaugh was assisted in the project by, among others, Charles F. Capen and Cpl. Bradford A. Smith of the Army Map Service, both of whom would assume important roles in planetary astronomy during the NASA years. No natural earth satellites were detected in the patrol, but Tombaugh was able to secure NSF funding for a program of regular planetary photography. He and Brad Smith stayed at New Mexico State University, developing an impressive photographic planetary patrol later supported by the Air Force and by NASA. A decade later, when NASA decided to establish a worldwide photographic planetary patrol, it was Tombaugh's patrol telescope design that was chosen as the standard.[49]

While Bernard Lovell's remarks concerning the influence of defense science quoted above have wide relevance to many specialties, they are especially apt in describing his own specialty, radio astronomy, and the particular branch of interest here, planetary radio and radar astronomy. R. H. Dicke and R. Beringer, using modified 1.25 cm radar equipment, first measured thermal radiation from the moon in 1946. That same year J. H. De Witt and E. K. Stodola of the U.S. Army Signal Corps succeeded in obtaining weak radar reflections from the moon using military 115 mhz equipment, and J. S. Hey and G. S. Stewart in Australia inaugurated their radar meteor research. The radar astronomy studies were to a large degree driven by the logic of the technique itself—the sensitivity and power of the equipment combined with the distances to various targets. The radar meteor research would expand rapidly, peak, and decline by the mid-1950s. The radar lunar work advanced throughout the 1950s, wherever suitable equipment was to be found. Radar astronomy promised more precise determination of the distances to the moon and planets, and hence a more accurate value for the astronomical unit. It also promised a means of mapping planetary surfaces and determining elevations and surface characteristics. It promised breakthroughs in the study of cloud-shrouded Venus, whose rotation rate and axis orientation were not known. The next target after the moon, Venus would not be successfully contacted until 1959. By then, the existence of more or less definite plans for spacecraft planetary exploration would lend an aspect of increased urgency and relevance to planetary radar astronomy.

The Navy's support of planetary research in the postwar period was perhaps less mission-oriented than that of either the Air Force or Army.[50] Office of Naval Research (ONR) support of university astronomy was varied, and included a number of planetary projects. The Navy's traditional mission of navigation and planetary ephemerides was one area in which the new computing techniques available after the war were implemented by celestial mechanicians such as Paul Herget, Gerald Clemence, Dirk Brouwer, and Wallace J. Eckert. In 1948 Clemence and Brouwer began a cooperative project between ONR, the U.S. Naval Observatory, Yale University, and the Watson Scientific Computing Laboratory to recompute ephemerides for the outer planets. The IBM Selective Sequence Electronic Calculator for the first time made the method of special perturbations feasible for calculating the orbits from the years 1653 to 2060 in forty-day intervals.[51]

Two other innovative areas of research also drew on the Navy's strong interests: radio-radar and balloon astronomy. Radar systems had been under intensive development at the Naval Research Laboratory (NRL) until 1941 when Vannevar Bush, head of the Office of Scientific Research and Development, "stripped control over the development of radar" from the NRL and "gave it to the newly established Radiation Laboratory at MIT."[52] In 1947 John P. Hagan introduced radio astronomy at NRL, and Cornell H. Mayer began work on radio emission from the sun. In 1956, a year after the accidental discovery of decametric radiation from Jupiter by Carnegie researchers K. L. Franklin and Bernard Burke, Mayer and his colleagues T. P. McCullough and R. M. Sloanaker turned to Venus, Mars, and Jupiter at 3.15 cm and acquired the first radiometric infrared measurements at other than optical wavelengths. The results seemed to confirm previous infrared radiometry of Mars and Jupiter. In the case of Venus, where the optical radiometry referred to the clouds and the radio results to the surface, their finding of a temperature of 600 degrees K set off an intense debate which was not really settled until the late 1960s, after multiple visits by spacecraft.[53] NRL radio astronomy groups, as well as several other teams, refined and expanded these observations.

The chief impediment to optical planetary astronomy and spectroscopy was the earth's atmosphere. The ONR became interested in planetary (and other) astronomy from high-altitude balloons when rugged and lightweight plastic fabrics became available, making stratospheric altitudes attainable. Attempts to observe a solar eclipse from unmanned Skyhook balloons in 1954, although only partially successful, were encouraging enough that ONR started developing a manned version, Strato-Lab.[54] Planetary spectroscopists were eager to get beyond both ends of the optical spectrum. However, the balloons were not able to gain enough altitude to get beyond the ultraviolet cutoff of atmospheric ozone. While sounding rockets were able to achieve such altitudes, stabilization and altitude control techniques were not far enough developed to make ultraviolet planetary spectroscopy viable.

The availability of military funding for basic science after World War II, the broad mission relevance enjoyed by such funding, and the increasing interest of the military in various astronomical topics, particularly the moon and planets, helped make planetary astronomy a more viable specialty than it had been before the war, when all such research had to be done as part of the research program of existing astronomical institutions. With the establishment of NASA as the only agency with a clear mission to engage in planetary exploration, the emphasis of much of this research shifted to the civilian agency. But the military, particularly the Air Force, continued to maintain an interest in and to support such research well into the 1960s.

AEROSPACE INDUSTRY

With the growth of the ballistic missile programs in the early 1950s the aircraft industry was in transition from building conventional aircraft (propeller planes) to commercial jets and military vehicles. By 1957 the major manufacturers of missiles and components had assembled scientific staffs, both of internal employees and outside consultants, to explore additional uses of the missiles they were developing for the Army and Air Force. In 1956, at General Dynamics, Krafft Ehricke was examining high energy upper stages (later to become Centaur) for the Atlas missile, which might be used for, among other things, "launchings of instrumented space probes to the lunar surface and into the inner solar system, primarily to Venus and Mars," and "establishment of a small manned orbital laboratory for a crew of three to inaugurate systematic preparations for deep space missions of manned spaceships."[55] John Naugle, at Convair's San Diego Scientific Research Laboratory, was asked to report to management on what scientific missions could be flown to Mars and Venus with an advanced Centaur upper stage for the Air Force's Atlas missile. At the same time, Oran W. Nicks at North American Aviation's Missile Division was conducting a corporate study for Mars missions using advanced ion propulsion, and was searching for consulting expertise in celestial mechanics.[56] Both Naugle and Nicks would later move into important positions in NASA headquarters, directing lunar and planetary programs. Similar activities were afoot at other aerospace companies, including the Space Technology Laboratories of the Ramo-Woolridge Corporation (later TRW) and Hughes Aircraft.[57]

Perhaps most significant of all the corporate efforts for the later scientific research on the planets was a series of colloquia arranged by North American Aviation's Missile Division. The Lunar and Planetary Exploration Colloquium met from May 1958 to May 1963, and grew out of the lunar base studies conducted for the Air Force "Strategic Interplanetary Systems" study mentioned above. The initial steering group for the colloquium included, among others, geophysicist Frank Press and astronomer Dinsmore Alter (both very interested in lunar studies) and RAND Corporation meteorologist William W. Kellogg, interested in comparative studies of planetary atmospheres, especially Mars and Venus. The scientific discussions at the colloquium included the current state of knowledge of the sun, moon, and inner planets, as well as the engineering and biomedical aspects of living on the moon and traveling to the nearer planets. In addition to astronomers, geophysicists, chemists, and military personnel, most major aerospace corporations were represented, and even Walt Disney Productions.[58]

As it became clearer that the United States would indeed have a program of lunar and planetary exploration, these ad hoc corporate groups became institutionalized as formal divisions, such as Douglas Aircraft's Planetary Meteorology Group, which independently published handbooks of the physical properties of Mars and Venus to serve as guides for engineers unfamiliar with the astronomical literature.[59]

McDougall has noted that in the aerospace industry, NASA and the Department of Defense constitute virtually the only buyers, providing both market and funds for research. "Hence, the industry must be an unabashed suitor of the state. . . . Firms place a premium on grantsmanship not unlike the way professors learned to hustle in the regime of largesse after Sputnik."[60] The intense competition for lunar and planetary contracts and the potential new market for advanced space propulsion and spacecraft systems required that such firms maintain in-house scientific expertise and be able to tap such expertise from the university community. One NASA official, commenting on a proposal from JPL for a summer institute in planetary science commented, "we would have to guard against participation in such an institute by large numbers of commercial employees and amateur astronomers who feel that this is a great way to get into the planetary game."[61]

The aerospace firms, eager to develop more advanced systems than could be justified for defense alone, often outpaced even the advanced plans of NASA's planetary programs. A propulsion engineer from Republic Aviation's Missile Systems Division wrote to NASA in May of 1961 that for two years he "had been conducting a feasibility study of a space vehicle for the reconnaissance of Jupiter and its satellites" and asked if NASA would be interested in funding more detailed studies. NASA replied that its own studies for the outer planets had not yet reached that level of detail.[62]

The commercial interests in lunar and planetary exploration generated more need for planning data on the moon and planets. In addition to the engineering information required to develop viable proposals (trajectories, environments of operation, timetables for flight, energy requirements), some indication of potential scientific observations which could be carried out by the instruments and the types of instruments suitable for such observations were required. To this end, commercial firms joined the growing number of parties interested in the current state of astronomical knowledge concerning the moon and planets, and entered the competition for the expertise and time of the relatively few competent lunar and planetary experts.

ASTRONOMERS AFTER THE WAR

Gerard Kuiper had trained at Leiden Observatory, receiving his Ph.D. in 1933. There he had developed an early interest in the planets, although most of his work from then until after the war was primarily on binary stars and white dwarfs. He had been recruited for Yerkes Observatory in 1936, after a year at Lick and at Harvard. Kuiper later attributed his binary star work to his interest in the solar system. He had reviewed a book on the origin of the solar system, and while he had become fascinated with the subject, he found the synthetic part of the book "entirely disappointing. I continued for many months to struggle with this problem and had to conclude that the state of Astronomy did not permit its solution. I was nevertheless fascinated by it. . . . I then determined to find a closely-related problem, that with finite effort would probably lend itself to a solution. This, I thought, was the problem of the origin of double stars."[63] Kuiper's publications show a majority of work before the war on double stars, with a continuing though much smaller proportion of planetary papers. After the war, however, his research returned almost entirely to the solar system.

Kuiper was sent to Europe twice as part of the ALSOS mission, a group of scientists and military personnel which a consortium of U.S. military groups sent to investigate the state of scientific war research, particularly in Germany.[64] While he did not return to Yerkes and full-time astronomy until late 1945, his research into European science was the occasion for making many contacts with European astronomers, especially the French planetary astronomers Bernard Lyot, Audouin Dollfus, and H. Camichel at Meudon.[65] During this time he became aware of the great scientific potential of the V-2, and also of advances in infrared detector technology that would open up new regions in the infrared wavelengths for planetary and other spectroscopy.

Shortly after the war Kuiper was back at Yerkes, and had initiated

FIG. 1.1. Gerard P. Kuiper. Photograph courtesy of NASA.

what would be one prong of an overall assault on the planets with every technique and every willing investigator he could manage. He made contact with a group at Northwestern University which included Robert J. Cashman and Wallace R. Wilson. Their collaborative work with Kuiper was the construction of an infrared spectrometer using a Cashman lead sulfide cell. The Cashman cell, which would remain the primary astronomical infrared detector well into the 1960s, allowed observations to be made between 1 and 2.5 microns, well beyond the photographic range even with sensitized emulsions.[66]

The new capabilities of the spectrometer called for new laboratory comparison spectra. Kuiper and Gerard Herzberg set up the spectrometer in Herzberg's spectroscopic laboratory at Yerkes. Herzberg overcame technical problems that had stymied earlier spectroscopists who attempted to duplicate planetary conditions in the laboratory.[67] The combination of the lead sulfide cell and Herzberg's innovations in laboratory spectroscopy opened the telescopic study of planetary atmospheres and surfaces to much more exact analysis than had previously been possible. Kuiper was quick to exploit these advantages, and did an enormous amount of spectroscopic as well as observational work in the years immediately following the war.[68] Indeed, numerous astronomers, like Kuiper's student Carl Sagan, contend that he was "the world's only full-time planetary astrophysicist" in the 1940s and 1950s.[69]

Kuiper became intensely interested in the potential of rocket exploration of the upper atmosphere, and shortly after his return to astronomy after the war had decided to make the focus of the fiftieth anniversary symposium at Yerkes the atmospheres of the earth and planets. He assembled an impressive program of speakers, "meteorologists, high-altitude specialists, and astronomers, in order that each group might learn of advances made by the others." Kuiper noted that he had been inspired by H. N. Russell's textbook *The Solar System and Its Origin*. Although many of the papers were reviews, all reported new results, and the conference was an enormous success. The proceedings, published in 1949, were sold out in a year, and an extensively revised edition was offered in 1952. The 1952 edition was reprinted as late as 1979.[70]

At the same time that Kuiper was systematically applying available astronomical techniques to the planets, he was also working on the subject that had gotten him interested in the solar system in the first place: cosmogony. After Russell had raised serious, ultimately fatal, objections to the encounter hypothesis in 1935, the origin of the solar system had no leading theory. But in the mid-1940s C. F. von Weiszsacker proposed a variant of the nebular hypothesis, in which turbulent vortices in the solar nebula formed, allowing the accretion of proto-planets and answering some of the original dynamic objections to the nebular hypothesis.[71] Weiszacker's theory was greeted enthusiastically by American astronomers, and once again the development of a solar system was seen to be a consequence of stellar formation rather than a freak accident. The significance of the difference between the two theories cannot be underestimated. For one thing, if the planetary system were formed as the result of an extremely improbably encounter of the sun with another star, then the planets represent mere debris left from the outer layers of the sun. If, however, planetary systems were commonly produced as a consequence of the general process of stellar formation, then characteristics of the planets and aspects of their origin deduced therefrom would have a direct bearing on stellar evolution.[72] Second, a major motivation for planetary study has always been the search for life beyond the earth. Under the encounter hypothesis the evolution of life would be one freak occurrence associated with a very improbable event, and the life on earth would probably be unique. On the other hand, if planets were commonly spun off accreting stars, then the prospects of extraterrestrial life become much greater.[73]

Kuiper's cosmogony, as it developed from the late 1940s, helped guide his research. The dynamics of the evolution of a planetary system from an accreting star led him to a systematic study of planetary satellites, asteroids, and other minor bodies of the solar system, particularly those whose orbital elements were peculiar, much as the nebular hypothesis had helped inform the research of Daniel Kirkwood a century earlier. He worked with one of his graduate students, Daniel Harris, on an enormous program of systematic photometry and colorimetry of the entire inventory of the solar system, starting in 1948, to acquire fundamental observations and synthesize them with reappraisals of the older observations. With another graduate student, Tom Gehrels, and in cooperation with Leiden Observatory, he initiated an ambitious asteroid survey to photographic magnitude 14.5 in 1949. His third graduate student in the 1950s was Sagan, whose four-part thesis examined possible indigenous organic matter on the moon and possible biological contamination by spacecraft, the radiation balance of Venus, and the production of organic molecules in planetary atmospheres.[74]

Kuiper also was well aware of the significance of the moon in such a cosmogony. The surface features and physical characteristics, the only ones on another body which could be studied in any detail, held powerful evidence on the dynamics and debris history of the early solar system. He conducted visual observations and photography from McDonald and Yerkes, and at times from Lick and Mount Wilson as well. When the Air Force became interested in lunar mapping in the later 1950s, Kuiper began a photographic lunar atlas project, and British lunar specialists Ewen Whitaker and D.W.G. Arthur joined Kuiper at Yerkes.[75]

Kuiper was president of Commission 16 of the International Astronomical Union, "Physical Studies of Planets and Satellites," for the entire decade of the 1950s. He took this opportunity to vigorously promote and coordinate worldwide various planetary and lunar studies. Prominent during these years were two major projects: University of Chicago–Air Force photographic lunar atlas, for which photographs were collected from around the world and an international patrol of Mars during the favorable perihelic oppositions of 1954–56.[76]

In addition to Kuiper, a number of other astronomers were also either continuing an old association with the planets or getting started. The perihelic opposition of Mars in 1956, the closest opposition since that of 1924, attracted quite a bit of interest in the planet. Kuiper's prior identification of carbon dioxide in the atmosphere, water ice in the polar caps, and the availability of a further infrared spectral range raised hopes for this opposition, at which detailed studies of locations on the disc would be more easily made. Better radiometers and the availability of radio telescopes held promise of better characterization of the energy balance of the planet. The availability of the Hale 200-inch telescope on Mount Palomar, opened in 1948, raised hopes for observations with much higher resolution than ever before.[77]

In 1953 Earl Slipher and A. G. Wilson organized an International Mars Committee at Lowell Observatory to coordinate U.S. observations; the Commission des Surfaces Planetaires did the same at Meudon, France, as did the Japanese Mars Committee in Tokyo. Seventeen observatories evenly spaced around the world in longitude cooperated. The National Geographic Society sponsored a Lowell Observatory expedition to Bloemfontein, South Africa.[78]

William M. Sinton and John Strong used the 200-inch telescope at Mount Palomar to perform a series of radiometric measurements. At Mount Wilson, Seth Nicholson and Robert S. Richardson used both the 60-inch and 100-inch telescopes for some visual, but mainly photographic, coverage with a motion picture camera used to catch moments of excellent seeing. Nicholson, who with Edison Pettit had performed the radiometric measurements of Mars at the 1924 opposition, was primarily a solar astronomer, but maintained an interest in planetary research. In 1938 he had discovered two additional satellites of Jupiter in a photographic survey with the 100-inch. Richardson, also a staff astronomer at Mount Wilson since the 1920s, had presented his "postwar plan for Mars" not in an astronomical journal but a science fiction magazine in 1944. He called attention to the "deplorably low state into which planetary observing had fallen," and proposed that by the next close opposition,

> we must be ready to assault the Red Planet with a high-powered battery of astrophysicists, electronic engineers, chemists, mathematical physicists, geologists, photographers, and opticians. These men should be selected, not only for their outstanding reputations, but for an open, flexible attitude of mind, as well as a demonstrated ability to work with others easily and without constraint. There should be close cooperation among them.[79]

In a popular exposition of the significance of the 1954 opposition, Richardson noted that interest in the planets had gone into decline around 1934, and that "on the whole the number of workers concentrating on the planets was insignificant compared with the number of those engaged in studying the stars. Even the close approaches of Mars in 1937 and 1939 passed almost unnoticed." But by the mid-1950s Richardson had detected "a strong revival of interest in the planets among astronomers" and "a new attitude in the public as well." Richardson attributed this change in attitude to developments in rocketry, and in particular singled out the February 1949 flight of the first two-stage rocket, a Bumper-Wac which reached the height of 250 miles. "Such achivements led some to believe that manned rocket flight to the moon and nearer planets might soon become a reality. . . . Engineers on rocket propulsion with whom I have talked have displayed a surprising interest in astronomy and interplanetary flight, an interest which would hardly be likely unless they thought there was a good chance of achieving it."[80]

The visual, photographic, and polarimetric studies at Meudon and Pic du Midi continued, with Dollfus succeeding his teacher Lyot. Camichel, and J. Focas from Athens, also worked at the Pic with Dollfus. For the 1954 opposition of Mars in May, Dollfus unsuccessfully attempted to detect Martian water vapor from a balloon more than 25,000 feet above Meudon, making photoelectric measurements centered on a narrow water absorption band at 8,250 A.[81]

In the years immediately following the Second World War a small number of astronomers turned their attention once again to the planets. To a large degree, these astronomers did their planetary work as a sideline, while depending on more mainstream astronomical topics for their standing within the professional community.

GEOPHYSICS AND OTHER DISCIPLINES

At the same time that Kuiper was making up for the time lost during the war, and setting out on his comprehensive program of planetary study, another scientist at the University of Chicago, Harold Urey, Kuiper's equal in vision, acuity, and scientific tenacity and passion, was also developing an intense interest in cosmogony and especially lunar studies. "Perhaps it will surprise readers that a physical chemist should undertake to prepare a book on *The Planets: Their Origin and Development*," Urey wrote in 1952, "and indeed it astonished me that I or anyone of similar training and experience should be able to say anything on the subject. As time has gone on my surprise has disappeared."[82]

Urey arrived at Chicago shortly after the war, in his fifties, and with a distinguished career to his credit. He had done his doctoral work under G. N. Lewis at Berkeley in the 1920s, studied quantum physics with Niels Bohr, developed a course and textbook at Johns Hopkins University on quantum mechanics and atomic and molecular spectroscopy, investigated nucleosynthesis in stars before the work of Bethe, received the 1934 Nobel Prize for his part in the discovery of deuterium, and led Columbia University's contribution to the Manhattan Project in production of heavy water and the gaseous diffusion method of producing uranium for the atomic bomb.[83]

Urey and geophysicist Harrison Brown had agreed to give a summer course in 1950 entitled "Chemistry in Nature." In preparing for the course in 1948 and 1949, Urey was introduced to current conceptions of the development of the earth, moon, and planets. Since the development of planetary bodies from the proto-solar nebula would have occurred at fairly low temperatures (as opposed to what was postulated by the encounter hypothesis, in which the planets would have formed from portions of the hot solar atmosphere), chemical forces would have dominated. Hence, as Urey wrote, "as astronomers have had undisputed possession of the field since ancient times, except for some interference from religious leaders and ancient religious writings, some discussion from other sciences should prove useful."[84]

Urey's basic cosmogony, initially, was very similar to Kuiper's, and the two of them worked together and exchanged ideas. Eventually, however, in the mid-1950s, their views became increasingly divergent, particularly in regard to lunar chronology and the nature of the protoplanets which were presumed to have formed as a prelude to planetary accretion.[85] But the fact that Urey and Kuiper, both distinguished scientists, began seriously working on solar system problems during the postwar period helped raise the status of planetary research.

The application of new geophysical and chemical techniques to problems in the earth sciences during the postwar decade, encouraged by the intensive and extensive investigations of the International Geophysical Year, resulted in an explosion of basic data about the earth's surface, interior, and atmosphere. New scientific disciplines and specialties grew up at the boundaries of old ones, as earth scientists struggled with ways in which to interpret and apply various data from specialties outside their own. The relationship of the earth to the solar system was probed by geologists and geochemists comparing terrestrial rocks and meteorites. It was being studied by geophysicists probing the impingement of the sun's radiation on the earth's atmosphere and ionosphere and the resulting phenomena in the magnetic fields and particles. The earth was beginning to be studied as a planet, and a number of scientists were beginning to look for comparisons between this planet and the others.

By the time travel to the moon and planets could be discussed in a serious and practical way, interest in those bodies had begun to extend far beyond the traditional disciplinary and institutional locus of astronomy. But while interest had gone far beyond astronomy, the means to study the moon and planets in the immediate future still lay with astronomers. And while some astronomers seemed to be taking greater interest in planetary study, the overall discipline had a full research agenda beyond the planetary system and a relative shortage of personnel and facilities to carry out that agenda. So long as planetary voyages were in the long-range planning stage, the study of the planetary system could proceed at its own pace. But once plans for lunar and planetary voyages began to be seriously developed, such basic knowledge would be more urgently needed.

Defining the Need (1958–1963)

The prospect of travel to the moon and planets held the imagination of the pioneers of spaceflight for many years. When automobiles were still a novelty, and with nothing more than a few overgrown skyrockets and some theoretical calculations to ground their dreams, the rocketeers of the early twentieth century foresaw with amazing clarity voyages to the moon and Mars. In like manner, when NASA was created just a few weeks after the first tiny U.S. satellite had struggled into orbit, planners incorporated planetary travel into its program. In planning and conducting satellite research, scientists and engineers could draw on decades of acquired experience and scientific study of the near-earth environment. Attempts to survey the field of lunar and planetary research, however, revealed a profound lack of basic information on these targets and their environments, information essential to the planning and conduct of planetary missions. Various groups and individuals, within and outside of NASA, eventually convinced the agency that it would have to enlarge its support for ground-based study of the moon and planets. Before the dreams of the pioneers could be realized, much remedial scientific work had to be done. But before the planners could be persuaded to support such work, they themselves had to be convinced that after so many years of study, astronomers had not mastered the nearest bodies in the sky.

SPUTNIK, SATELLITES, AND THE PLANETS

During the 1950s a substantial amount of effort had been put into research in the near-space environment. Balloons, sounding rockets, and high altitude aircraft all probed the upper reaches of the atmosphere, at the boundary of space. Earth-orbiting satellites had been seriously discussed and planned since immediately after World War II, and as part of the International Geophysical Year (IGY, 1957–1958) both the Soviet Union and the United States planned to launch small artificial satellites into orbit. At various symposia of a more visionary nature travel to the moon and planets was taking on a more concrete and serious tone, as rocket engineers began to develop the various systems which put travel beyond earth orbit on the verge of feasibility. Indeed, in retrospect it can be said that there were no serious engineering or scientific impediments to the development of satellites and space probes which could not have been overcome in the late 1940s. What was lacking before Sputnik was the commitment of money by government (the only party able to finance such ventures) and an organization able to marshal, organize, and direct the various scientific and engineering resources toward the goal of spaceflight.[1]

The launch of Sputnik on October 4, 1957, the response of the United States, and the counterresponse of the Soviet Union provided the impetus to shift earth-orbiting satellites from experiments to routine operations and interplanetary flight from vision to reality. Although President Eisenhower initially saw no need for changing the IGY satellite plans or accelerating the space program, the public response to Sputniks 1 and 2, the clamor among the military agencies for a much expanded space program, and the series of partisan congressional hearings criticizing the president's management of space activities all helped push him toward a greater response. On the recommendation of his newly reconstituted President's Science Advisory Committee (PSAC), Eisenhower proposed to Congress in April 1958 the creation of a new agency to manage all civilian space activities.

NASA had a very broad mission to conduct scientific research in space. The lack of definition in the legislation which created the agency gave NASA managers a wide latitude in how they went about creating a space science program. The urgency felt by the American people and their politicians to get on with the space program, and the relatively few groups and individuals with experience in conducting research using space techniques, had an important effect on the character of space science at NASA. It meant that the agency would have to rely on existing competence and interest wherever possible.

Historian R. Cargill Hall has introduced a useful characterization of the "branches of the space science tree": (a) sky science—physics and chemistry of the upper atmosphere, including particles and fields and the interaction of the sun and earth atmosphere; (b) planetary science—the geophysical exploration of the bodies in the solar system—essentially earth science writ large; (c) life science—including two separate areas, space medicine and the search for extraterrestrial life; (d) space astronomy—taking advantage of the space environment to overcome the detrimental effects of the earth's atmosphere on astronomical observations. Hall characterizes the sky scientists as a "cohesive group, well-organized and reasonably well publicized." Sky scientists had been active in upper-air rocket and satellite research since the close of World War II, had developed a variety of instruments for their research, and were experienced in the construction and use of such instruments in the rather severe and specialized environment in which they had to operate. It is not surprising, therefore, that "it was largely individuals with such sky-science backgrounds, linked organizationally, familiar with each another, and experienced in sounding rocket research, who first came to occupy space science positions in NASA. . . . all things considered, sky science held the upper hand in the space agency."[2]

Planetary scientists, on the other hand, were point for point almost the opposite. Planetary science depended heavily on images—be it the optical observation of other planets through large telescopes or the geological classification of landforms from aerial surveys or topographic maps; and there was only a most rudimentary technology for returning pictures from space. Some spectrographs and other radiation detectors capable of being pointed at a particular region of the sky had been developed for sounding rockets and automated balloons, but the accuracy of pointing controls in early space vehicles constrained the usefulness of such instruments. (Indeed, balloons had been chosen by the pioneers in airborne planetary spectroscopy precisely because their instrumentation was too heavy for sounding rockets and required a more stable platform.) The laboratory techniques that geophysicists and geochemists used to analyze and date rocks and meteorites could not easily be translated into instrument packages weighing only a few pounds. In any case, how were such instruments to be transported to the planets when engineers were still struggling to get only a few pounds into orbit or perhaps to the moon? As Hall writes, "If in 1959 and 1960 obtaining a sufficiently powerful and reliable launch vehicle occupied NASA as a whole, acquiring effective leverage in the agency's space research program was a major challenge for the nation's planetary, especially lunar, scientists."[3]

The "effective leverage" for the nation's lunar and planetary scientists came from two different directions within the NASA organization during 1958 and 1959. At headquarters, Robert Jastrow became a champion of lunar exploration, while the Jet Propulsion Laboratory chose to make planetary (and lunar) exploration its own specialty. Jastrow had worked with NASA Space Science chief Homer Newell in the Naval Research Laboratory's upper atmosphere research group, and had transferred with Newell and others to NASA in October 1958. Jastrow, a theoretical physicist, was to set up a Theoretical Division (for the yet to be built Goddard Spaceflight Center) to conduct research in all fields of physical science that could be investigated from spacecraft. In the course of doing some background reading, Jastrow came across Harold Urey's 1952 book *The Planets,* and was so impressed that he decided to fly to LaJolla, where Urey was professor-at-large. Urey talked with Jastrow, explaining his theories of the importance of the moon as a "rosetta stone of the solar system," that the moon was probably geologically dead, that it preserved on its surface and in its interior a record of conditions in the early solar system, and may even be one of the early planetesmals from which the planets formed. Jastrow "resolved to bring Harold's fascinating message to the people in NASA," and arranged for Urey to give lectures at NASA headquarters on January 15 and 16, 1959. The timing was perhaps fortuitous, for on January 4 the Soviet *Luna 1* passed within about 5,000 km of the moon, carrying various particles and fields instruments, once again achieving a space first at the expense of the United States.[4] After the lectures, and a discussion with Newell, Jastrow and Urey drafted a memorandum proposing that an intensive program directed to a lunar soft landing in 1961 be instituted.[5]

Newell later said that the "Ranger program was in effect born on that day," although "no specific spacecraft was planned, no schedule drawn, no engineering details worked out."[6] The schedule, engineering details, and selection of scientific experiments soon followed. An ad hoc Working Group on Lunar Exploration was formed, chaired by Jastrow, and including Urey, James Arnold, Frank Press, and Harrison Brown (all chemists or geophysicists). The Lunar Science Group (the group had several names) by July had added Von Braun's colleague Ernst Stuhlinger, geophysicists

FIG. 2.1. Homer E. Newell. Photograph courtesy of NASA.

Maurice Ewing and Gordon McDonald, physicist Bruno Rossi, astronomer Thomas Gold, biologist Joshua Lederberg, and JPL's space science chief Al Hibbs. Its function was to provide NASA "with close contact with eminent scientists interested in lunar science, and conversely to provide the scientific community with an assured contact with NASA," to assist the NASA Office of Space Sciences in selecting experiments to be carried by lunar vehicles, and to furnish general advice in the lunar science program.[7]

Throughout 1959 Newell was staffing the Headquarters Office of Space Sciences with new NASA employees, and also assembling working groups of outside consultants in areas other than lunar science. The working groups, as well as the various new positions at headquarters, went through many permutations of title and job description as Newell and his staff grappled with the problem of just how to carve up the scientific applications of space techniques.[8]

The space science staff was constrained by the fact that they were subsumed under Spaceflight Development, and as such had to organize for some purposes in terms of the vehicles to be used. Earth-orbiting satellites used smaller rockets, needed less elaborate guidance, and were in general less complex than spacecraft destined for lunar or planetary distances. But Newell and his staff were anxious to keep close to established scientific disciplines wherever possible, and so used one organizing principle for budgetary purposes, another for long-range planning, and yet another for the organization of the office.[9] But whether the lines were drawn on the basis of vehicles, office procedure, or scientific discipline, planetary studies were always separate from astronomy. As such, when it came to considering ground-based supporting research, planetary astronomy tended to fall in the cracks.[10] Moreover, with the location of space sciences under spaceflight development in the early years, there was a tendency to think of the scientific aspects of the program solely as "packages" to be carried on vehicles. With such an attitude, and with the perception of the agency's mission as limited to space techniques, ground-based supporting research tended to be thought of as outside the agency's purview.[11] In any case, with Jastrow's lunar group in a prominent and active role, and continuing problems in the development of JPL's intermediate-energy planetary stage, Vega, headquarters inclined more and more throughout 1959 to favor lunar probes almost to the exclusion of planetary. Several ten-year plans drafted during the latter part of 1959 describe lunar exploration as the main emphasis and defer an attempt at even a "Venus near miss" until 1967.[12]

By December, however, NASA had worked out the transfer of the Von Braun team and their launch vehicle programs, including the powerful Saturn. The disclosure by the Air Force earlier in the year of a hitherto secret, more powerful Agena B upper stage for the Atlas, together with the acquisition by NASA in July 1960 of the uncompleted Air Force Centaur upper stage for the Atlas, meant that a vehicle capable of sending significant payloads to Mars and Venus would be available in just a few years.[13] In light of these new developments, NASA officials traveled to JPL in December 1960 to work out details of the lunar and planetary plans for the next several years.[14]

JPL engineers had a long tradition of seeking challenging assignments. They had toyed with the notion of space travel and satellites during their missile work in the late 1940s and through the 1950s. With a number of successful space projects under their belts, they were already, before NASA's creation, looking beyond the moon. "[W]e wanted a good challenge, and that was *the* technical challenge, getting a useful payload to a planet. It was really tops in engineering challenge—propulsion, guidance, communications, you name it."[15] While discussions were going on in Washington toward setting up a civilian space agency, there was a lot of talk at JPL about what role the lab should play within the new organization. They quickly decided on the lunar and planetary portion, partially because "nobody else wanted it," and partially because of the challenge.[16] While JPL had an extremely able and creative engineering staff, there were few "pure" scientists at the lab. When they decided to prepare for lunar and planetary exploration, word got around to Ray Newburn, who had been working as a mathematician on the attenuation of radio signals in rocket exhaust. Newburn had studied astronomy as a graduate student at the California Institute of Technology before taking a summer job at JPL where he stayed on. Newburn suggested to Robert Megrheblian, his section chief at the lab, that with all the talk about lunar and planetary missions perhaps it would be worthwhile to put some ideas on paper of the kind of science that could be done from probes. In the spring and summer of 1958, Newburn, together with Marcia Neugebauer, searched the astronomical literature and interviewed astronomers at Mount Wilson and elsewhere. They drafted a document summarizing the state of knowledge of the solar system, suggesting areas of research that might be amenable to space probes.[17]

Meanwhile, through the summer JPL had been cooperating with Von Braun's team on the Pioneer lunar probes and Director William Pickering had been campaigning to have JPL become "the national space laboratory" for NASA. By September, Pickering had requested studies of planetary payloads which would use Von Braun's Juno IV. In October, with an informal agreement worked out to make JPL a part of NASA, a major study effort began at JPL to develop a five-year plan for lunar and planetary exploration.[18]

Newburn and Neugebauer joined the study team, and a revision of their exploratory paper circulated among various astronomers. Hibbs, Newburn, Neugebauer, and others traveled to interview Urey and other scientists with planetary interests, and the group compiled a report that became the first comprehensive proposal for exploration of the solar system.[19] In it, the specifics of celestial mechanics, vehicle designs, navigation, communications, tracking, power supply, and other matters were worked out in detail. A program for five years was suggested, taking advantage of every launch opportunity for expeditions to Venus and Mars, and filling in the slack times with lunar missions. But in a section on scientific considerations, where plans for investigating various solar system bodies were described, a complementary program of ground-based planetary astronomy was proposed in support of the probe program:

> During the past half-century, only study of the stars has received a great deal of attention by the professional astronomer. . . . Astronomers, lost in the depths of interstellar and extragalactic space, have never returned to the planets. . . . It is self-evident that no probe study is scientifically justifiable if the same technical results can be obtained with terrestrial equipment, nor should any space exploration be undertaken without the fullest support of simultaneous earth-based research. . . . This report on the proposed space-probe program suggests such a supporting research program because of general agreement that supporting research, costing only a fraction of one percent of the total space research funding, could well spell the difference between immediate success and a discouraging stretch-out in the overall space program.[20]

The plan which JPL presented to NASA in May showed that the experts in Pasadena had firmly decided on the planets as their goal, with the moon occupying a subsidiary role. Moreover, they had begun to build up their scientific staff. At first, Hibbs later recalled, the JPL workers had to learn just what the scientific issues were in lunar and planetary research. They had chosen the goal for the engineering challenge and the absence of competition. They then tried to learn the scientific and astronomical aspects: "Well, we read the books. We read the British Patrick Moore's book on Venus, and we grew our own astronomers at JPL to read the books about Astronomy. And we asked Gerard Kuiper to help us. We got Gerard de Vaucouleurs . . . to come and give us lectures." This was just a preliminary step, to get an introductory perspective on what kinds of scientific observations could be done from the probes. For actual design and development of hardware, JPL knew it would have to expand its scientific staff.[21]

JPL had decided to join NASA with the intention of planning, developing, and executing the entire deep-space program. They had interpreted informal discussions, during the summer and fall of 1958 which led to the five-year plan as "a commission for JPL to plan a long range space program for NASA," and believed that the study would "result in NASA's *major space program* but would not incorporate the entire program."[22] Hence they were distressed at the sudden appearance of Jastrow's lunar group at headquarters, and at subsequent decisions to focus on lunar exploration and to cancel Venus and Mars probes which had been scheduled in the earlier planning document.[23]

Headquarters' lunar and planetary staff met on December 16, 1959, to "establish the lunar and planetary program objectives for the next decade," and decide on various divisions of responsibility, schedules, payloads, and funding. They assigned responsibility for lunar and planetary spacecraft to JPL, and asserted that for the immediate future the moon would have priority over the planets.[24] Meanwhile, Pickering had written to Newell's superior at the Office of Space Science, proposing that the charter of Jastrow's lunar group be expanded to include planetary and interplanetary science, that scientists with such expertise be added to the group, and, more important, that the chairmanship of the group revert to Hibbs, at JPL, instead of Jastrow at headquarters. When headquarters and JPL officials met on December 28 Pickering drew attention to the difference of opinion between JPL and headquarters regarding whether planetary or lunar work should be emphasized. The NASA officials stood firm in their intention to focus on the moon, but "pointed out that the planetary work should get underway at once, and that there would be a planetary try every time the near planets, Mars, and Venus, were in optimum position for a planetary mission."[25]

Newell then met privately with Pickering, Hibbs, and Frank Goddard to clarify policy matters between JPL and headquarters. On the matter of lunar and planetary planning, Newell asserted that overall program planning was a headquarters activity, with the detailed engineering planning and execution a responsibility of JPL. As for Jastrow's lunar committee, Newell pointed out that steps had already been taken to combine lunar and planetary exploration into one committee, and that Newell himself would chair it, with membership drawn from NASA and its centers, and scientists acting in a consultant and advisory capacity.[26] Thus in a relative standoff, NASA headquarters had asserted its lead role in program planning and management, while JPL received its commission to specialize in lunar and planetary exploration. It was, however, an unstable truce that would occasionally erupt into conflict over the following decades, and color the JPL–headquarters negotiations over planetary astronomy.

Where there once had been been none, a lunar *and* planetary program had finally emerged in early 1960, even though specific missions, schedules, and targets would continue to change in response to various factors, from launch vehicle troubles to congressional oversight. In the course of planning for lunar and planetary exploration, JPL, headquarters, National Academy of Sciences Space Science Board, and other groups sought out knowledgeable people to advise them on the current understanding of the moon and planets. In the course of these discussions it was discovered that there was a lot of work that still could be done from the ground, but as yet had not been. A widespread network of alliances had formed between engineers, scientists, administrators, and institutions, for whom planetary exploration, as well as planetary astronomy, would be helpful in the pursuit of their several different goals. One of the forums that enabled these diverse groups to discover their common concerns was arranged by an aerospace company seeking to expand its business from missiles to scientific launch vehicles, and from ballistic trajectories to interplanetary ones.[27]

LUNAR AND PLANETARY EXPLORATION
COLLOQUIUM

In the course of conducting lunar base studies during 1956 and 1957, the Missile Division of North American Aviation began studying the potential of the moon as "the springboard for future solar system exploration." North American decided to arrange a series of informal colloquia, which lasted from May 1958 to May 1963, and afterward published the proceedings, including transcripts of discussion.[28] In the first year, over four hundred people attended; the following list shows the relative proportion of participants, from most to least numerous:

> engineers, astronomers, physicists, space medicine researchers;
> geologists, biologists, geophysicists, geochemists;
> scientific writers and editors, and officers of the armed services and government agencies;
> chemists, astrophysicists, meteorologists and aerodynamicists;
> oceanographers, seismologists, radiochemists, mathematicians, astrodynamicists, cosmologists, zoologists, and photogrammetrists.[29]

The first two colloquia, in May and July of 1958, were devoted exclusively to the moon, with presentations by Dinsmore Alter of the Griffith Observatory, Frank Press of Caltech, and Harry Vestine of the RAND Corporation, among others. At the July colloquium, A. G. Wilson, who had recently moved from Lowell Observatory to RAND, presented a penetrating and thoughtful analysis of the problems of planning space exploration. Wilson first recalled the "pre-space era" when space enthusiasts "rivaled Madison Avenue in their contributions to the science of motivation research," and when "almost every type of probable and improbable gimmick was used as a lure to cast before the military and civilian fund-dispensers in order to make a sale." Now that space exploration as an activity seemed securely established, it was time to sort out motives and confront "the problem of arriving at a unified, scientifically responsible program, or syntax, of space exploration." Taking the exploration of the earth as a model, Wilson went on to outline steps in the exploration of a new region. The first steps, he noted, were the assembling of "all existing preknowledge which can be obtained by indirect methods," and then extension of that knowledge for "specific exploration purposes," from which comes the design of the first exploratory efforts. In the case of the moon, Wilson observed, "the astronomer has not been concerned with landing a space ship on the moon and has not made all of the observations or performed any of the simulation experiments which would be useful for this purpose. This is an operation which is only now being proposed or conducted by nonastronomical research organizations."[30]

This theme reverberated through the colloquia for years. At the March

1959 meeting James Edson, also formerly with the Lowell Observatory, and subsequently assistant to the director of advanced research and developmemt of the Army, reported on observations of Venus near inferior conjunction that he and others had made in the 1930s. Since their group was broken up by the war, the data had not been reduced or evaluated. Edson showed how such data could be used to arrive at models of the vertical structure and composition of the Venus atmosphere. Urey questioned Edson after his presentation, finally asking "Why can't we get somebody to measure and analyze these things [already existing data]?" adding that it was inconsistent to spend enormous sums of money for probes while ignoring "the handy little devices that lie on our doorstep." Hibbs, at that time engaged in a recruiting effort to build up JPL's planetary and astronomical expertise, added, "We would be perfectly willing to hire any astronomer who wants to take over your photographs and reduce whatever data we can get out of them—if there is one who can stop looking at the galaxies long enough to do the job. There's money available; there's interest available. Where are the people who know what they are doing?" Edson said that perhaps the interested ones were the geophysicists, but they would have to learn astronomical techniques. Urey suggested, "Surely the techniques are not too esoteric to be learned. Maybe you should forget the astronomers and try the physicists."[31]

The theme of the need for ground-based supporting research and the conservatism of astronomers was picked up at the September 1959 colloquium by S. M. Greenfield, a RAND meteorologist serving as an Air Force science advisor. Greenfield's task was to introduce, and then recapitulate and appraise, three synthesizing sessions in astrobiology, earth sciences, and astronomy. In introducing the colloquium, Greenfield first noted that sciences determine the directions of their research "on the basis of their curiosity about observed phenomena; they do not concern themselves with seeking answers to assigned problems." But, Greenfield continued, the boundaries between applied and pure research were beginning to blur with space techniques. "I believe that now there has arisen a rather anomalous situation: the existence of an assignment for the pure researcher. There are recognized gaps in our basic knowledge of the solar system and these must be filled before the science of astronautics can proceed in a logical manner." Whereas previous colloquium discussions had centered on what opportunities astronautics could supply to various sciences, the question for the present colloquium was "What can these sciences do, here on the Earth's surface, for astronautics? . . . What research programs can extend our knowledge in these sciences—for example, planetary physics— which can be done on the ground but which for some reason—lack of funds, marginal returns of the effort, allure of more exciting research areas— have been neglected to date?" Greenfield finished his introduction by noting that NASA and other agencies in Washington were extremely interested in this subject, and were looking for ways to give a logical direction to their programs. Although the colloquium was not a formal organization, its proceedings were read with interest in Washington, he said, and this fact made the present colloquium perhaps the most significant to date.[32]

In the keynote speech for the astronomy sessions, Wilson observed, "astronomers have been about the most conservative group with regard to space and only a few have shown interest in its possibilities. There now exist large, enthusiastic groups in space medicine and space law. I suppose we'll have to wait until there's a flourishing discipline of space dentistry before the astronomers will get excited." Wilson then went on to point up the difference between the needs of the space engineers and the results of sciences which had, quite appropriately, been choosing problems to work on and solve without regard to their possible astronautical utility. Engineers, he related, wanted to know the value of the astronomical unit to six or seven significant figures, and when the astronomers could only produce three the engineers ask, "what have you astronomers been doing for two hundred years?" He then went on to suggest how the new developments in space had altered the astronomical agenda. As he saw it, the first task (95% of their time) of astronomers was to continue to do what they had been doing, working on problems deemed most important by the community. But for the rest of the time, Wilson suggested assisting the space engineers to remedy the "ignorance of near [solar system] space," and also attempting to exploit the new possibilities opened up by spaceflight for all kinds of astronomy. He suggested some specific projects, but above all argued for achieving a balance of effort and money between ground- and space-based research, estimating the annual research budget for the whole world in astronomy to be roughly equivalent to the budget for one probe mission. He also pointed out that the expanded space activity called for training more scientists, and that the ability to do interdisciplinary research would be of paramount importance.[33]

After two days of papers, discussion turned to a recapitulation of the conference. Zdenek Kopal, who had been organizing the Air Force lunar mapping project in Manchester (U.K.) and at Pic du Midi, noted that most of the topographic information about the moon "goes back to the nineteenth century," and Kopal did not think that the data were reliable enough to specify the location of a point with an accuracy of five miles. What was worse, his research had shown that much of the data on heights and locations of lunar craters and features had often been inaccurately copied and recopied from previous sources. Kopal noted that in Kuiper's forthcoming critical selection of lunar photographs in the Chicago Lunar Atlas, for only 10% of them was the time the photograph was taken known to within 15 minutes, making them useless for deriving altitudes from their shadows. Kopal pleaded for doing all possible from the earth before sending probes to the moon.[34]

Gerard de Vaucouleurs noted that astronomers "cannot do everything at the same time, so they do what is most interesting from the view of physics and astronomy. That is why they look at the stars and galaxies."

But if astronomers were given some direction as to what planetary information is most important to space planners, they could at least have a criterion of selection for projects. While acknowledging the need, de Vaucouleurs warned that "the approach must be right. You cannot simply advertise in the papers and get astronomers to leave their observatories or the universities, and take a job in the government and industry."[35] The total number of astronomers of any kind was so small, de Vaucouleurs warned, that it would be better to give money to people already working in the field, "because once they have joined you they are out of astronomy."

Wilson pointed out the need to make extensive and intensive planetary observations, not just spatially, but in time resolution as well. "This has not been done, not because of limitations interposed by the Earth's atmosphere, but because of limitations of budgets and the availability of suitable instruments. . . . What is required are instruments of moderate size distributed around the world, which would be available for full-time work on the planets, and even more important, people to do the work. There are few qualified or interested astronomers for this purpose."[36]

Celestial mechanician Samuel Herrick discussed the desiderata in his field, noting that Gauss, Laplace, and Lagrange had only "solved the problems that nature presented to them, but they didn't solve the new problems that artificial satellites are now presenting to us." Basic constants of the solar system were poorly determined. The value of the solar parallax, "the ratio between the heliocentric system of units and the laboratory units used by the engineer in his launching of rockets," had taken on new importance now that planetary voyages were planned.[37]

Greenfield then summarized the session, pointing out that every governmental agency involved in space flight was represented, and that, although the colloquium had no official status, its deliberations were followed with interest. Greenfield's summary, although not to be realized in a significant way for some years, represented a clear statement of the rationale that would drive NASA's planetary astronomy program:

> A logical approach to space exploration demands the full utilization of a parallel program of Earth-based research. By providing a basic foundation of knowledge, such a program would allow us to exploit fully the information presently available or attainable from, and on, the Earth, and leave for the space probe experiments the data that can be obtained in no other way. . . . information gathered from such probes can be interpreted much better in an existing context of reliable basic knowledge. Feedback of probe-acquired data may then suggest continued Earth-based experiments.[38]

Thus in September 1959, many of the major desiderata and problems of ground-based planetary astronomy in support of the planetary exploration program had been aired. Nearly every astronomer who had even shown a casual interest in planetary research appeared at the colloquia at some time or other, and many attended faithfully. Through this medium, JPL space science staff, some headquarters officials, and scientists and engineers from all sorts of disciplines and organizations had a chance to become apprised of the planetary astronomy research front.

SPACE SCIENCE BOARD

Similar sentiments were being expressed in other forums, and in many cases by the same individuals. The Space Science Board (SSB) of the National Academy of Sciences grew out of the Technical Panel on the Earth Satellite Program of the U. S. Committee on the International Geophysical Year. In June 1958, even before NASA had been officially authorized by Congress, the SSB convened and began to solicit proposals for scientific experiments to be conducted in space, which the board members would evaluate through various committees. Although the SSB was to be strictly an advisory body, the status of its recommendations to NASA, and the issue of NASA's responsiveness to those recommendations, made for what Newell called a "love-hate" relationship, in which the SSB "served as watchdog for the scientific community." The SSB served as a kind of scientific "shadow government," and although NASA emerged from an initial skirmish firmly in control of the civilian space program, the force of the National Academy of Sciences behind the SSB gave its recommendations some considerable weight.[39]

Leo Goldberg's Committee on Astronomy of the SSB, in a January 14, 1959 interim report, recognized the partial overlap of its interest with that of other SSB committees, and recognized planetary astronomy as within its purview. The report noted the utility of balloons for infrared spectroscopy of planetary atmospheres, and suggested astronomical satellites for ultraviolet planetary photometry and radio astronomy of Jupiter below the ionospheric cutoff.[40] In November 1959 E. R. Dyer, Jr., of the Space Science Board Secretariat, collected several documents in which "several contributors have arrived at substantially the same conclusion concerning the necessity for making the maximum possible use of what information we now have about planetary physics and concerning the desirability of more fully exploiting earth-bound techniques in support of future planetary probes." On November 13, 1959, Dyer sent the documents to the members of Urey's Committee 1, Chemistry of Space and Exploration of the Moon and Planets; Alan H. Shapley's Committee 7, Ionospheres of the Earth and Planets; Keffer Hartline's Committee 11, Psychological and Biological Research; the Panel on Extraterrestrial Life of the joint Armed Forces–National Research Council (NRC) Committee on Bio-Astronautics; and others.[41]

The first of the documents consisted of "notes for long-range planning" of Shapley's Committee 7, Ionospheres of the Earth and Planets, from a meeting on September 17–18, 1959, which identified two groups of investigators with common interests but only partial communication: ionospheric physicists familiar with the latest theoretical interpretations of the earth's atmosphere and ionosphere but little knowledge of the latest observational results on other planets, and planetary astronomers whose situation was just the opposite.

The second document was an excerpt from the Panel on Extraterrestrial Life. The panel encouraged the formation of institutes for the study of the solar system at academic institutions having appropriate expertise, and suggested that contract support should be provided these institutes "so that an adequate degree of stability and continuity will be established from the beginning." Noting that telescope time for planetary studies was inadequate for the work proposed, the panel recommended that "an observing station should be established with telescopic and auxiliary equipment to be used for planetary studies. The operation of this Solar System Observatory, which would be built in a place chosen especially for its favorable observing conditions, would be the prime responsibility of one group, but its facilities must be available to all Solar System investigators." This suggestion would lead to serious Air Force plans to establish its own planetary observatory near Cloudcroft, New Mexico.[42]

The third item was a paper by James B. Edson of Army Ordnance, "More Payoff per Dollar from Interplanetary Probes (A Void and How It May Be Filled)." Edson first noted the relatively modest cost of earth-based planetary observations and their importance to the scientific return from probes. He then stated that planetary physics "constitutes almost a void" when compared to other astronomical fields, because astronomers possess the techniques and instruments, but not the geophysical training and interests, and geophysicists possess the latter but not the former.

Edson proposed close coordination between the ground-based and probe programs, use of balloon-borne and earth satellite-borne instruments, plus a coordinated network of radio astronomical observations. He urged that more astronomers be trained in physical planetology, since "in this country today there are probably not more than a dozen living men who have made original research contributions to physical planetology, and probably no more than half of them are now working full time in this field," Appendices to the memorandum listed various observations and estimates of the present state of knowledge, chance for improvement with earth-based techniques, and chance for improvement with probes, as well as an estimated budget. Edson's plan would, in a revised form, become a blueprint for a later NASA program.[43]

The Space Science Board received a number of replies to Dyer's memo and enclosures. Among the several positive respondents was Roman K. C. Johns of Baird-Atomic, who sent the SSB's Lloyd Berkner a copy of a similar memorandum he had submitted to the Committee on Geodesy on March 9. Johns proposed to Berkner that the SSB sponsor a conference of astronomers and geophysicists interested in planetary research, adding that he had discussed the matter with Edson, S. M. Greenfield, de Vaucouleurs, Aden Meinel, and others. There were also negative replies, some arguing that ground-based planetary astronomy would soon be superseded by space probe results, and was therefore not worth the effort.[44]

The first explicit discussion of planetary ground-based supporting research and technology appears in the SSB's minutes for the fifth meeting, May 7–9, 1959, where it was simply noted that there were certain gaps in laboratory research which needed to be filled.[45] In a report to the President's Scientific Advisory Committee in November 1959, the SSB called attention to the fact that "researches to develop our knowledge of what to expect in space and planetary programs" were required, but that "a crack exists between NSF and NASA into which falls the longer-range researches essential to advanced space projects of the future."[46]

In March 1960, Urey's committee heard Edson's briefing on his proposed network of planetary observatories and endorsed his recommendations in another interim report to the SSB.[47]

As Berkner had promised Johns, the SSB discussed the topic of ground-based astronomical observations of the planets at an all-day and evening conference on planetary atmospheres held June 24. The conference itself grew out of discussions that Shapley, L. A. Manning, and H. G. Booker (all of Shapley's SSB Committee on Ionospheres of the Earth and Planets) had with Urey. As a result of the discussions, Shapley's committee recommended a network of ground-based observatories to "provide a continuous planetary patrol," decided to arrange the June conference, and recommended changing its own name to Atmospheres of the Earth and Planets.[48] The June conference impressed the board concerning what could be learned about the planets from ground-based astronomy. Urey's committee presented the following resolution, which was adopted by the SSB without objection:

> The Space Science Board of the National Academy of Sciences recognizes that the national program for the exploration of the moon and planets can benefit greatly from the energetic application of the techniques of ground-based astronomy. At the present time, the field of lunar and planetary astronomy suffers from inadequate equipment and insufficient personnel. These can be remedied with the help of funds representing a small fraction of the total budget for lunar and planetary space-probe exploration.
>
> Considering the great scientific value of such observations to the national space research program, the Board strongly recommends that appropriate governmental agencies consider providing adequate support and encouragement to scientists and institutions already available for expanding this program.
>
> Such investigations should be regarded as an integral part of the space program.

NASA Deputy Administrator Hugh Dryden, NSF Astronomy Program Officer Geoffrey Keller, and others were present for this meeting. Dryden suggested that this resolution be passed on to Presidential Science Advisor George Kistiakowski "for consideration . . . as a part of national science policy."[49]

The conference and subsequent resolution started a chain of events which ultimately led to NASA's initiating a significant program of just the sort that various scientists and JPL had been proposing. But it did not happen immediately. The planetary atmospheres conference findings concerning ground-based planetary astronomy seems to have impressed the Space Science Board and others more than it impressed NASA. Dryden's rather cool reaction to the resolution above was an early indication. In addition, relationships between the SSB and NASA were not very good at this time, particularly this month. The previous November, Berkner had written a long letter criticizing numerous aspects of NASA's handling of space science, and had directed the letter over NASA's head to Kistiakowski, the President's Science Advisor. Newell later wrote, "The going was difficult, and criticism continued until in June 1960 the author felt compelled to put out a workpaper on the subject."[50]

In a letter to Shapley, Manning communicated his "vivid impression" gained from the SSB-sponsored conference on planetary atmospheres. "It seems extremely important to me that every effort be made to increase the amount of work done in planetary observations before the need arises to instrument planetary launchings," he wrote. "High quality scientific experiments do not grow in a vacuum. They appear in the minds of people who have been asking questions about their subject for a period of time and who have been actively working in the field." Acknowledging that ground-based planetary research is much cheaper (an argument often used by Urey), Manning thought that a stronger argument for supporting ground-based planetary research was that "only in this way can we get maximum participation of the scientific community in planetary research. Once the scientists can be hooked to the problem, they will automatically want to verify their hypotheses by probe experiments." Manning's letter was circulated to all the participants in the planetary atmospheres conference and to the SSB. Newell, however, remained skeptical, commenting on his copy that "this sounds like a very poor argument," and found Manning's assertions that NASA had a responsibility to support such ground-based planetary research "questionable."[51] Later, Newell would become convinced that just such an approach was necessary.

Meteorologist William Kellogg wrote to Berkner about the SSB's Arcadia Planetary Atmospheres Conference, noting that several members of the "JPL team . . . did not miss a session, and took an active part in the discussions." He also noted a "very strong consensus on the matter of support for earth-based observations of the planets." After discussing the planned program of planetary probes and the expense of the program, Kellogg noted that it had come to the SSB's attention repeatedly that "we are not providing for the essential backup program of earth-based observations required to interpret the probe observations." Kellogg suggested three general areas for support: (a) a cooperative program of planetary observations from observatories around the world, patterned after the International Geophysical Year, which "could lead to an International Planetary Year"; (b) more support for small balloons and aircraft as astronomical platforms; and (c) expanded support for planetary radio and radar astronomy.[52]

At the end of March 1961, Berkner wrote to James E. Webb. the new NASA administrator under the Kennedy administration, enclosing two major policy statements—one on the role of man in space and the other on the support of basic research. In the latter, the SSB argued that "first, the fullest advantage should be taken of research opportunities on the ground and by means of balloons and rockets which valuably augment satellite and space probe scientific investigations. . . . Ground-based instruments and observatories can provide many scientific answers in studies of the upper atmosphere and significant new data concerning the Sun, the Moon, and the planets." The SSB noted that the objectives of the space program in the 1970s would focus on the solar system, and that to that end basic research in this area must begin in the 1960s. While the NSF's "organic" mission was to support such research, it had not done so because of administrative difficulties. The SSB recommended that NASA provide expanded and stable support for university researchers, support which would be separately identified in the NASA budget to protect it from being ignored. The board further recommended that "an expanded program of earth-based observations and research by means of balloons and rockets should be pressed to exhaust the scientific potentialities of these approaches. In parallel context, instruments and techniques for planetary observations should be tested in the Earth's environment in a preliminary way."[53]

Two years after the extensive discussions at the Lunar and Planetary Exploration Colloquium, three years after JPL's initial attempts to prepare for lunar and planetary exploration, the SSB was firmly convinced of the need for a major program of ground-based planetary astronomy, and a cadre of interdisciplinary scientists interested in the planets was forming. During the same period, similar events and discussions were occurring at NASA headquarters. Again, it was largely the same cast of characters bringing to bear the same retinue of arguments and evidence in favor of a program of ground-based planetary astronomy.

SPACE SCIENCES STEERING COMMITTEE

In the spring of 1960, while the SSB was preparing for the June Planetary Atmospheres symposium, NASA headquarters was revamping the organization of the Space Sciences Office and its advisory structure. NASA created a separate Office of Lunar and Planetary Programs, and in place of ad hoc committees such as Jastrow's lunar group, created a Space Sciences Steering Committee, with Newell as chairman, to provide advice to top NASA officials for selection of scientific experiments for various missions. This committee was in turn advised by various discipline subcommittees, composed of headquarters and centers representatives, and with individual scientists as consultants.[54]

Nancy Roman chaired the Astronomy Subcommittee. As JPL representative, Ray Newburn was selected. Newburn had joined the Space Sciences

Division at JPL, contributed to the 1959 five-year plan, participated in the Lunar and Planetary Exploration Colloquia, and was in the process of organizing and recruiting astronomical expertise at JPL. He wasted no time in bringing the planetary astronomy message to the subcommittee.

At their first meeting, April 25, 1960, "Newburn pointed out that a JPL Committee on planetary probes had recommended continuous observations of all planets."[55] At the second meeting Newburn gave a lengthy presentation on the types of ground-based, balloon-borne, and possible earth satellite planetary astronomy needed to support the flight program. Newburn's handwritten notes show that he discussed synoptic monitoring of Mars and Venus ("Can't overemphasize"), spectrophotometry of atmospheres ("successful design of soft landing vehicles will depend [on this]"),

FIG. 2.2. Nancy G. Roman. Photograph courtesy of NASA.

and infrared spectrophotometry from balloons and aircraft (U-2). Reflectance spectrometry of surfaces, Newburn thought, would be difficult because of a lack of laboratory comparison spectra, and in any case spacecraft would probably sample at least Mars before reflectance spectra could be interpreted faithfully.

At the subcommittee's third meeting, July 21 and 22, 1960 (after the SSB Planetary Atmospheres Conference in June), "ground-based planetary work was discussed at some length although it was realized this is not the specific area of this committee." The subcommittee also discussed a memorandum which Gerard Kuiper had presented to the Planetary and Interplanetary Sciences Subcommittee early in June. The Astronomy Subcommittee noted an "over-emphasis on visual observations from the ground," and concluded that "a program of planetary work from above the earth's atmosphere is vital," but came to no decision concerning ground-based work, and postponed the subject until a future meeting "when more information is available."[56]

At the fourth meeting, October 24–25, 1960, which Newburn could not attend, the minutes record that "planets *per se* are not this subcommittee's domain, but as astronomers we deal with them." The subcommittee did think a number of planetary investigations worthy of study (primarily spectroscopy) and discussed planetary observations that could be carried out from astronomy satellites.[57]

The Astronomy Subcommittee spent most of its time deliberating on the desiderata for astronomy satellites. Plans for an orbiting telescope were advancing at headquarters, and the subcommittee did not concern itself too much with discussion of any kind of ground-based astronomy—planetary or stellar—except at Newburn's urging.

But in another subcommittee, more directly concerned with the planets, and chaired by Newell himself, similar arguments were being put concerning the need for additional ground-based planetary astronomy. The Planetary and Interplanetary Sciences (P&I) Subcommittee included as members Al Hibbs and Richard Davies from JPL (both of whom were active in the JPL efforts to solicit astronomical cooperation in the planetary program), Gerhardt Schilling, a geophysicist and headquarters astronomy program chief (Schilling would leave NASA before the year was out to become Associate Head of the Planetary Sciences Department at RAND), and Robert Jastrow, among others. The subcommittee included as consultants Kuiper, Joseph W. Chamberlain of Yerkes, Joshua Lederberg, Edward Ney of the University of Minnesota, and Carnegie Institute of Technology's Philip Abelson.

At their second meeting, June 7, 1960, JPL was assigned to "investigate the need for a ground-based, balloon, and satellite support program for the Lunar and Planetary Explorations," and "committee consultants [were] asked to consider the study of the earth as a planet, and the use of research on the ground or in balloons, sounding rockets, and earth satellites, as a precursor to planetary research in deep space vehicles."[58] Kuiper brought up the subject of planetary astronomy, which was discussed at length by the P & I Subcommittee. "The committee was in complete agreement that a ground-based supporting research program was a vital part of the overall plan for Lunar and Planetary Explorations."[59]

Newell sent Kuiper's written version to the SSB for the use of its study group on planetary atmospheres, and solicited comments. Copies of the memorandum were distributed to other subcommittees of the Space Sciences Steering Committee for discussion.[60]

Kuiper began by stating the problem: are the present and anticipated astronomical facilities sufficient to support the space program? Kuiper pointed out that training in astronomy does not fully prepare one for planetary astronomy, which requires some knowledge of geophysics and other disciplines, as well as specialized astronomical techniques. Kuiper estimated that for the past twenty years perhaps one Ph.D. in "planetology" was trained every three years.[61]

He proceeded to outline the major scientific work that needed to be undertaken: discovery of other minor bodies in the solar system, studies of the composition and structure of planetary and satellite atmospheres, measurements of fundamental quantities (mass, diameter, etc.), surface mapping, studies of surface texture and composition, various studies of natural satellites, investigation of the interplanetary medium, and laboratory and theoretical studies needed to interpret observational findings. Kuiper then discussed which of the many desiderata could best be achieved from spacecraft and which offered promise of solution from the ground.[62]

The fifth section of the memorandum attempted to show by example that "the inclusion of ground-based observations in an over-all program is not merely a matter of economy; it seems a logical necessity, by the nature of the problems themselves," because without the ground-based program, "we would get data, but no integrated science."[63] Kuiper argued that the list of problems cited showed that a dedicated lunar and planetary observatory, "an institution not now existing in the West," would aid NASA in achieving its goals, enhance and increase the scientific output of the planetary program, and provide leadership in an area now of great interest but hitherto neglected "because it falls between established disciplines (geophysics and astronomy), is not part of the University corricula [sic], and requires facilities quite beyond the powers of Universities, even if supported by modest research grants."[64]

The proposed facility, Kuiper continued, would need a scientific staff of broad and diverse background, a spectroscopic laboratory with long path-length tubes, a geophysical laboratory including facilities for comparative polarization studies, cartographic facilities, telescopic focal instruments, photographic laboratories, measuring and photogrammetric equipment, and a library. Radio telescopes would not be part of the proposed facility, because of their expense, the existence of adequate national facilities, and the modest requirements for observing time for planetary radio astronomy. "It is concluded that the vital radio data will be forthcoming without NASA entering the field."[65]

Optical telescopes, however, were another matter. Since they were privately owned, and heavily booked with stellar programs, their use for planetary astronomy could not be assured. Kuiper saw "no sufficient reason to duplicate the 200-inch telescope," because only a portion of the needed planetary research required such an aperture, and the Palomar instrument had been made available for certain specific programs. As for existing U. S. telescopes in the 36 to 100- inch aperture range, these were fully occupied with stellar and galactic research. The Kitt Peak 36-inch was booked eighteen months in advance; the Mount Wilson 60- and 100-inch telescopes were "regularly booked full" except for a few nights around full moon for the 60-inch; the new 120-inch at Lick was "now fully occupied by the California astronomers on stellar programs"; and the 82-inch McDonald telescope was under heavy competition from Chicago, Texas, and Indiana astronomers.[66]

Kuiper concluded by proposing that a 60-inch reflector, at a "first rate location (presumably a mountain site in the Western U.S. or possibly Hawaii)," plus perhaps an additional 24 to 36-inch instrument be part of the proposed overall facility.

Kuiper wrote Newell on June 28 with a copy of the memorandum and a letter from Jesse Greenstein attesting to the crowded observing schedule of the Mount Wilson telescopes. Kuiper added that his Lunar and Planetary Laboratory (LPL) in Tucson, now under construction, had only a 36-inch telescope and that the addition of a 60-inch would allow the laboratory to be much more productive and also train new students. He suggested a cooperative agreement between NASA, the University of Arizona, and LPL.[67]

Newell received evaluations of Kuiper's proposal from several quarters. Most acknowledged the need for more ground-based planetary astronomy, but they disagreed on how to go about achieving this.[68] Newell had also asked Geoffrey Keller and Gerald Mulders at NSF to read the memorandum, and asked them to comment on what NASA's role should be. Newell learned that NSF definitely had no plans to support such a facility. However, he also learned of the Air Force interest in building a lunar and planetary observatory at Cloudcroft, New Mexico.[69]

In a report circulated to Newell among others, Nancy Roman wrote that she had accompanied the working group on the Cloudcroft facility to its proposed site, and had "participated in a lengthy discussion on the need for a planetary facility," which "strengthened but did not alter my opinions about such an observatory." Roman wrote, first, that she agreed that such an observatory would be a "useful addition to American astronomical facilities," and that "no large telescopes exist on which an appreciable amount of time can be devoted to planetary research." She believed that this was not the result of prejudice, but merely that the pressure from other programs was so great that "planetary programs must take their share of time and not more." Roman thought that a request from a "respected scientist" to use telescopes for "serious scientific studies would be favorably received." She then specified the type of telescope which would be suitable for such an observatory, basically in agreement with Kuiper's memorandum. On the topic of visual observations, however, she felt them to be "of dubious value," as would be a world-wide network of such patrol type telescopes. "The five astronomers with whom I have discussed a planetary observatory, are in unanimous agreement on the subject of visual planetary observations. If anything, they feel more strongly than I, that it is inadvisable to devote an appreciable fraction of time and effort at a new planetary observatory to visual observations."[70]

As of late 1960, then, Newell found himself in a serious predicament. He was first of all committed to a rational, carefully planned, and scientifically sound program. Although NASA headquarters would maintain control over appropriate aspects of the program, such as selection of scientific instruments, scheduling and planning of missions, etc., Newell firmly believed that the diverse scientific community in the universities and elsewhere must be integrated into the planning and execution. He further believed that space research was not a passing fancy, but would endure and prosper in future years. With this perspective, he wished to make sure that a wide and strong base was built in the various communities outside NASA. If it were true that outside intervention into the existing system of astronomical research was required to assure the success of the lunar and planetary spaceflight program, then he was prepared to do what was necessary to insure that the requisite supporting research be done. On the other hand, he believed that scientific participation in the space program had to come from the "grass roots level," and was wary of trying to get scientists to do things they were reluctant or opposed to doing. Newell was becoming increasingly convinced that astronomers were just not interested in the planets, and seemed to believe that the future of planetary research lay with geophysics.[71]

The responses Newell got from various parties and the discussions in the Astronomy Subcommittee did not give a clear indication of the extent to which planetary astronomy really needed to be stimulated. And it remained ambiguous as to whether the problem was a lack of interest, lack of personnel, or lack of instruments and facilities.

At the third meeting of the P & I Subcommittee, on September 7, 1960, the results of various discussions were summarized. There was "sharp disagreement" over whether ground-based methods were capable of giving "new results worth the labor necessary to inaugurate such a program," and all agreed that "the program should not be allowed to degenerate into mere monitoring of the planets." Among the supporters of planetary astronomy, there was also disagreement over the relative merits of ground-based and balloon-borne work. The feasibility of establishing a new observatory was questioned; according to one member, "if this is necessary to begin a space probe program, then it is already too late." The Air Force plans for Cloudcroft were unclear, and needed investigation. Finally, "many questions remain concerning management, operation, funding, timing, etc."[72]

The Astronomy Subcommittee was asked by the P & I Subcommittee for "suggestions and ideas as to planetary astronomical research that NASA should include in its program."[73] At its fifth meeting, February 23–24, 1961 at JPL, the P & I Subcommittee heard a presentation by Roman on space available for planetary experiments on the three planned Orbiting Astronomical Observatories. Roman also described areas of research, such as Martin Schwarzchild's *Stratoscope* balloon-borne telescope, which her office was supporting. She distributed a summary of comments from members of her subcommittee suggesting "needed planetary research."[74]

The Astronomy Subcommittee held a three-day meeting at JPL March 22–24, 1961, part of which was spent working out sections for the NASA long-range plan. The subcommittee broke up into "sub-subcommittees," one of which considered "lunar, planetary, and interplanetary astronomy," and was made up of members Newburn, Dirk Brouwer, R. T. Jones (Ames Research Laboratory), and John O'Keefe (Goddard).[75]

At the subcommittee meeting at JPL, a new report was made available summarizing the deliberations of JPL's committee on planetary astronomy, which had been meeting since February of 1960. The report was written by J. Allen Hynek of Northwestern University, and included contributions by Edson, de Vaucouleurs, Greenfield, Kopal, Kuiper, A. G. Wilson, as well as JPL scientists J. Goldsen and J. Kitchen. The 26-page report contained many of the features of Edson's prior memorandum, augmented with references to the various committees of the Space Science Board (particularly the ad hoc committee on planetary atmospheres which had been meeting at JPL while the report was being written). The report discussed the status of planetary studies among astronomers, and quoted a letter from Fred Whipple to Frank K. Edmondson: "For the more than thirty years that I have been in astronomy, the professional astronomer has considered planetary and interplanetary research as generally *infra dig.* . . . those astronomers who were interested in the subject received practically no support, financial or moral, until after World War II, when the military became quite interested in the upper atmosphere and mildly interested in the space beyond."[76]

The report noted the "formidable atmospheric barrier" which had frustrated attempts at visual planetary astronomy, the present overall scarcity of astronomers (in a random walk one would be one hundred times more likely to meet a bank president than an astronomer), and the overenthusiasm of amateurs who filled the professional vacuum in planetary astronomy. The committee argued for a partnership of astronomers and geophysicists, and for the extensive use of balloon-borne observations to achieve spectral ranges and optical definition unattainable from the earth.[77]

Specific recommendations were given for: (1) establishing a worldwide planetary patrol, by financial and other support to observatories selected strategically with respect to longitude; (2) a high-altitude facility from which balloons could be launched on a regular basis to make planetary observations; (3) a data reduction center, perhaps with a telescopic facility attached; (4) a systematic comparative survey of astronomical seeing at the sites of the cooperating observatories.[78]

In a following section, specific programs of lunar and planetary observation were suggested. Systematic lunar mapping (coordinated with the existing Army and Air Force projects), a thorough program of radiometry of the lunar surface, and a photographic patrol of the lunar surface to detect recently suspected short-term changes on the moon were advocated. Extensive lunar observations were argued to be "essential to the appropriate and adequate interpretation of telemetered information" from probes.[79]

The committee began the section on planetary observations by noting that while the United States "may not reap the political and scientific prestige of priority in lunar exploration, the situation may be quite the reverse in planetary exploration." Balloon observations, being above much of the atmosphere, offered the best promise for preliminary maps of the planets to be visited, while the proposed worldwide patrol network would be important for tracking meteorological phenomena on Mars. The committee elaborated the importance of most classes of possible planetary astronomical observations, showing a clear preference for balloon-borne observations to remedy problems associated with atmospheric absorption and scintillation.[80]

A survey of twenty-six observatories around the world which would be "potential members of the lunar-planetary observing network" was presented, as was an expanded version of Edson's proposed standard observing kits for cooperating stations. The standardized kits would allow photographs in various filters, low dispersion spectra, photometry, radiometry, and polarimetry to be made in rapid succession within a five to fifteen-minute period.[81]

The various subcommittee minutes are silent about the reception of this report and its proposals. However, in October Roman wrote a lengthy critique which circulated through the Office of Space Sciences. Roman's evaluation would be expected to carry considerable weight. She was the resident professional astronomer in the office and the main liaison with the professional astronomical community. Roman was "in complete agreement" with the notion that more information could be extracted from ground-based studies and that balloon observations were most promising. However, she disagreed with "the implication that there are large numbers of competent personnel who might be attracted to such a program and that the only lack at present is sufficient monetary support," and was "not at all in sympathy with the detailed recommendations of this report."[82] She then went through the report, page by page, offering a detailed critique of various recommendations.[83]

Roman had many specific disagreements with the proposed program, but in the end her criticisms were more about the approach to the problem than the need or value of ground-based planetary astronomy. She conceded the value, and to a more limited degree the promise, of ground-based and balloon-borne planetary astronomy. But JPL seemed to offer a solution that would be a large "crash" project, organized centrally from the top down. Roman favored rather a more traditional approach of soliciting specific proposals for specific projects from recognized, competent investigators. In the end, she recommended "that JPL be requested to outline the specific problems which should be attacked rather than generalities. When these have been listed in order of priority JPL, Headquarters, and outside consultants can consider them and formulate a sensible program leading to their solution."[84]

The advice Newell and others were getting from many quarters converged on several points. There was an apparent need for more ground-based planetary astronomy. There was a shortage of telescope time for planetary work, and considerable crowding in observatories generally.[85] There were few trained planetary astronomers. But several obstacles to implementing a solution remained. First, planetary astronomy fell in the crack between astronomy and geophysics, and some institutional solution to this ambiguity had to be found, since the geophysicists tended to think in terms of *in situ* measurements, and the astronomers seemed to be genuinely unconcerned with planetary work. This would eventually be solved by creating an administrative niche and budget item at headquarters for planetary astronomy. Second, planetary astronomy fell in the crack between ground- and space-based research, between NASA and NSF. This problem would eventually be solved by high-level policy agreements specifying the degree to which NASA had an interest in, and could properly become involved in, ground-based astronomy generally. Third, it was still unclear just how valuable and in what specific ways planetary astronomy could assist the probe program. The answer to this question would not follow with sufficient force to spur action for almost two years, when a ground-based redetermination of the Martian surface pressure would throw Mars landing capsule plans into a tailspin. Fourth, the planetary program took a back seat to the lunar program, and NASA's initial involvements in ground-based astronomy were in the areas of lunar studies, mainly mapping and radiometry. With President Kennedy's decision to make a manned lunar landing the primary goal of the space race, lunar work acquired even higher priority. Finally, the awareness of some kind of Air Force interest in constructing a planetary observatory at Cloudcroft raised the possibility that NASA might again be caught in the dilemma of having invested heavily in a project, only to have it cancelled because of duplication of facilities.

CLOUDCROFT—THE AIR FORCE LUNAR AND PLANETARY OBSERVATORY

The Air Force mission interest in the moon and planets, its style of funding basic science, and the Sacramento Peak solar observatory were discussed in chapter one. As part of an expansion of the Sacramento Peak facilities in 1959, a solar furnace was planned and land acquired nearby. The solar furnace plan was subsequently abandoned, and suggestions were solicited for the use of the Cloudcroft site. John Salisbury, head of the Lunar and Planetary Exploration Branch of the Air Force Cambridge Research Laboratory, put in a bid for a lunar and planetary observatory. Air Force officials considered this proposal the most promising, and commissioned an observational evaluation of the site.[86]

When Nancy Roman traveled to Sacramento Peak on an information-gathering trip, she discussed the need for a planetary observatory with the Cloudcroft group. S. M. Greenfield and Gerard de Vaucouleurs were doing a preliminary survey of the Cloudcroft site at the time (May 1960). Between December 1960 and September 1961 a further survey was carried out under a contract to New Mexico State University. Clyde Tombaugh was the project supervisor, Brad Smith did most of the data reduction, and Charles F. Capen was the chief observer. Smith and Capen had assisted Tombaugh during his search for natural earth satellites.[87]

The initial report, in December 1961, found the Cloudcroft site's average (mean) seeing (per five-minute interval) to be slightly inferior to Kitt Peak. However, since Cloudcroft was to be a planetary observatory, and planetary photography involves short exposures during instants of crisp seeing, the survey contractors attempted to compare the average best (modal) seeing per five-minute interval. On this criterion they found the Cloudcroft site to be superior to Kitt Peak.[88] Donald Menzel, who had initially helped establish Sacramento Peak and conducted that site survey, was contracted by the Air Force to assemble a committee to evaluate Cloudcroft's suitability on the basis of the site survey, and also advise "on the most effective way of establishing a planetary program in the best overall interests of the nation." On February 20–23 the committee toured the Cloudcroft site by air and on foot, discussed the methods of the site survey with Tombaugh and Smith, inspected the instruments and methods used, and held several meetings to arrive at a conclusion.[89] The Menzel committee unanimously concluded that "Report AFCRL 1064 does not represent an adequate survey for the selection of a site for such an observatory. However, the contractors had fulfilled their contract and had done about as well as one could with the survey under the several limitations imposed by higher authority." The primary disagreement the Menzel committee had with the 1961 version of Report 1064 was the definition of astronomical seeing, and the comparability of Kitt Peak and Cloudcroft. There was disagreement over how one goes about quantifying the notion of astronomical seeing, and whether stellar standards should be the same as planetary standards. However, the committee agreed unanimously that the observations made did not support the alleged superiority of seeing at Cloudcroft over that at Kitt Peak for planetary work, and the conclusions of Report 1064 were revised.[90]

The Menzel committee's deliberations further revealed that there were certain political restrictons on Air Force support for the observatory. In the first executive session Menzel asked one Air Force representative present if a committee decision that more study was needed would jeopardize the funds for the project. The official could not answer immediately, but telephoned Dr. Millsaps at the Air Force Office of Scientific Research and replied: "There is a verbal agreement that if the site were declared favorable, money would be made available for the installation. On the other hand, if the determination is unfavorable, it is anybody's guess where the money will come from. Dr. Millsaps advised the committee not to consider the financial questions and stick strictly to the scientific merits."[91] Smith reported that "the Air Force put rather severe restrictions of time and location on our work," and Salisbury, replying to Menzel's question as to why Sacramento Peak was not considered for the site, replied, "This was a ground rule under which we operated, handed down from higher levels." Salisbury further reported that his group was not given special funds for the site survey, and had to take $53,000 from their research budget. The "original directive read 'as soon as possible' in very strong terms. I heard this verbally interpreted as 'get it done in a year.'"[92]

The Menzel committee, while finding the Cloudcroft site to be "not only imperfect but relatively poor," nonetheless heartily endorsed the notion that a lunar and planetary observatory was needed:

> The committee would reemphasize, however, that they were very favorably impressed by Lt. Salisbury's presentation of the Air Force research program of the Lunar and Planetary Exploration Branch. . . . The committee unanimously feels that the Air Force program as outlined needs and deserves the type of ground-based support that a large planetary observatory would give.
> The committee therefore strongly recommends that the funds originally allotted to this program be held in reserve for the establishment of a planetary observatory at some more favorable location.[93]

The fact that the funds were apparently tied to the Otero County site and that a decision of the committee against that site might jeopardize the entire observatory troubled the committee, since they endorsed the scientific program and perceived the need for such an observatory. Menzel and Salisbury continued to try to find a way of establishing an Air Force lunar and planetary observatory at a scientifically acceptable site, but resistance from higher up within in the Air Force command persisted.[94] Four years after the initial reconaissance of the Cloudcroft site, it was the Air Force that was faced with duplicating existing or projected facilities. For in the interim various circumstances had led NASA to conclude it could wait no longer in more vigorously supporting lunar and planetary astronomy. When the Air Force was finally ready to move ahead with its solar system observatory, NASA already had several telescopes under contract, and other programs in place.[95]

During 1960–61 it was apparent that the Air Force was committed to a major lunar and planetary observatory. This, together with other factors already discussed, helps explain NASA's reticence in moving to involve itself in ground-based planetary astronomy in a significant way. With the findings of the Menzel committee, whose report was distributed in May

1962, Newell and Roman's cautious attitude in approaching the scientific community was reinforced. If NASA could learn one thing from the Cloudcroft incident at the time, it was how *not* to go about remedying the perceived lack of facilities and personnel for lunar and planetary research.

THE AIR FORCE CAMBRIDGE RESEARCH LABORATORY PROGRAM

The proposed Cloudcroft Lunar and Planetary Observatory was only part of the Air Force Cambridge Research Laboratory program. Since the establishment of the NASA-JPL lunar and planetary exploration program, most of the Air Force had resigned itself to less than an operational role on solar system bodies other than the earth. But it still wished to maintain competence in the area, and the idea of a lunar base was still alive, if more discreet. At the Menzel committee hearings, John Salisbury characterized the Air Force program as designed to "complement the NASA program rather than duplicate it," and described it as comprising three parts: ground-based, balloon, and rocket observations, backed up by a program of laboratory spectroscopy and theoretical analysis.[96]

The Air Force saw itself in need of its own observatory for the volume of ground-based work it contemplated. At the time of the Menzel committee meetings, various combinations of optical telescopes (80, 60, and 40 inches) were being considered, along with an infrared collector of 600–800 feet in diameter. The Air Force had already provided a 40-inch reflector to Pic du Midi for the lunar mapping program, and had contracted a great number of optical and infrared studies of the moon, Mars, and Venus. But the biggest problem it foresaw was staffing the new observatory with competent personnel.[97]

While specific plans for Cloudcroft seemed rather vague and contingent upon selection of Otero County as the site of the facility, the general program under development by the Air Force was broad in scope and seemed to encompass most available techniques for studying the moon and at least the inner planets. So long as the Cloudcroft facility remained alive as a possibility, NASA ran the risk of being caught in duplication of another program. Even if the Cloudcroft facility were to be built, however, it would not solve the problem of the shortage of trained planetary astronomers to staff it, and indeed might aggravate the problem.

THE JPL OPTICAL ASTRONOMY GROUP

JPL was getting impatient with the seeming lack of action on what was to them an already demonstrated need for facilities and expertise. JPL continued to press for implementation of the recommendations of the 1961 Report 33-37. The Space Sciences Division at JPL began to attract more astronomers and other scientists, and an impressive, aggressive group formed at the lab. Along with the general buildup of scientific expertise at the lab, JPL made special efforts to attract both established and new Ph.D.s, especially astronomers, as part-time consultants. Various difficulties JPL was already having with the scientific community regarding control of experiments submitted for spacecraft missions, combined with resentment over the tactics they were using to build up the lab's scientific expertise, must also be set within the context of the criticism directed at NASA as a whole that the agency was draining the nation's scientific and engineering personnel and facilities for what was a goal of dubious scientific worth.

Reinhard Beer, a new Ph.D. from Manchester, joined JPL in 1960, interested in infrared spectroscopy. Hyron Spinrad, having received his Ph.D. from Berkeley, joined the optical astronomy group in 1961. He immediately began looking for large-telescope time to do near infrared planetary spectroscopy, particularly of Mars and Venus. Ronald A. Schorn, a new Ph.D. from the University of Illinois trained in millimeter-wavelength radio astronomy and radiometry joined the group in 1962.

JPL astronomers were not as successful as they would have liked at gaining large-telescope time, particularly at Mount Wilson and Palomar. This led to somewhat strained relations between these facilities and JPL, and to a sustained effort on the part of the JPL optical astronomy group to get their own telescopic facilities.[98] But JPL was unable to obtain any large facility without headquarters' and congressional approval. As an interim step, Ray Newburn got approval in late 1961 to acquire a used Nishaimura 16-inch reflector, and after acquiring an unused site from the Smithsonian Astrophysical Observatory, Newburn and the JPL optical astronomy group established the Table Mountain Observatory in the fall of 1962. Charles Capen, of Tombaugh's New Mexico State Group, moved his family onto the mountain and began a visual and photographic planetary patrol.[99]

The JPL group immediately began planning for larger facilities. A 10-foot millimeter-wave radio dish was erected, and construction was started on a support for a 16-foot dish. By the fall of 1963, geologists and civil engineers had surveyed the Table Mountain site and pronounced it safe for a larger telescope. Newburn and Spinrad had begun design studies of a 48-inch reflector and coudé spectrograph combination for the observatory. But no approval for a larger telescope came from headquarters.[100]

NASA headquarters followd the approaches recommended by Roman in her review. During this period, between the time NASA had become convinced that something had to be done to gather more ground-based information on the moon and planets, but before it had decided to construct major telescopes and facilities, it attempted to obtain some of the information it required from already interested investigators and existing facilities.

KUIPER AND THE LUNAR AND PLANETARY LABORATORY

Meanwhile, Kuiper had also decided that he could not wait for any individual government agency to begin to fill in the gaps in the basic knowledge of the solar system. As his "Need for a Ground Based Lunar and Planetary Observatory" memorandum shows, Kuiper had been making as much use as possible of Mount Wilson, Yerkes, and McDonald, but his work and that of his students was demanding more and more telescope time, time which was getting increasingly difficult to procure. In the fall of 1960, shortly after Kuiper's memorandum, while NASA was deliberating, JPL was researching memorandum 33-37, and the Space Science Board committees were working on their studies, Kuiper was already attempting to establish, on his own, just the kind of organization he had outlined.

From Chicago he moved the lunar mapping project and staff, his graduate students, the major editorial projects, and a number of his collaborators to the University of Arizona, where he established the Lunar and Planetary Laboratory, whose purpose was to

> create in a favorable academic setting a research and teaching unit concerned with the study of the moon and planets. Favorable has reference to (a) the presence of research scientists in supporting fields, such as meteorology, geology, geochemistry, physics, and electrical engineering; (b) clear skies, low humidity, accessibility to modern telescopes and electronic computers; (c) proximity to national laboratories and industrial facilities; (d) a well-developed or developing graduate school; (e) proximity to interesting geological terrain features for comparative studies of moon and earth.[101]

By early 1962 LPL had five senior staff members (Kuiper, Harold L. Johnson, Aden Meinel, Tom Gehrels, Stuart Hoenig), five research associates, twelve research assistants, and nine support staff members. Kuiper had assembled support for the various programs and staff from almost every astronomy funding agency available—NASA, NSF, Navy, and Air Force—mainly through continuation and expansion of support he had already been receiving for various projects.[102]

Kuiper and his assistants rebuilt an infrared grating spectrometer acquired from another department at Arizona, and adapted it to the Texas 82-inch telescope. Parts of the infrared spectrometer Kuiper and Cashman had built shortly after the war were recycled into a ten-channel automatic recording spectrometer for faint sources. For laboratory comparisons, he borrowed one of Gerard Herzberg's multiple-reflection absorption tubes while the LPL's 72-foot, 8-inch diameter tube was being installed. He sent Tobias Owen, a graduate student who had studied with Kuiper at Chicago, to the Institute of Astrophysics at Ottawa to receive instructions from Herzberg on setting up the tube. Harold Johnson's digitized photometer was moved to LPL from McDonald. For the asteroid program, Kuiper worked out cooperative arrangements with Jan Oort at Leiden Observatory, where Dr. and Mrs. C. J. Van Houtens blinked and measured plates pending the construction of LPL's own blink comparitor. Paul Herget at the University of Cincinnati carried out the orbit calculations for new discoveries, and LPL cooperated with the asteroid program at Indiana University, where Gehrels had previously spent one year.

For telescope facilities, Kuiper made use of arrangements with Arizona's

Steward Observatory, with Kitt Peak, McDonald, and others. At LPL Johnson designed a 21-inch cassegrain telescope. A 29-inch telescope was jointly funded by ONR and NSF for photometry of the thousand brightest stars, and some years later was modified for planetary photography. Kuiper had designs ready for a 60-inch optical telescope, with provisions for all specialized instruments needed for planetary astronomy. He was also considering an infrared "light bucket," between 10- and 28-foot aperture, based on designs of Meinel and Johnson.[103]

Kuiper was moving at the same pace he had established at Yerkes after the war. In addition to his work in founding the Lunar and Planetary Laboratory, his ongoing research programs, his continuing site surveys, and his participation in multiple advisory committees, Kuiper had space flight projects to think about as well. In October 1961 he had been selected as principal investigator of a team which would interpret the photographs that the Ranger spacecraft sent back as they headed toward lunar impact. (The coinvestigators were geologist Eugene Shoemaker and Harold Urey). Kuiper was also appointed as an experimenter in the Surveyor lunar soft-landing project. An internal NASA memorandum of December 1961 expresses interest in a proposal of Kuiper's to develop a planetary entry capsule, and also some interest in the infrared telescope mentioned above. But the memo notes, "we should proceed with caution, for though Kuiper's group is growing steadily, the possibility for over extension must be avoided."[104]

Kuiper's laboratory was just the sort of institution he had outlined in his June 1960 memorandum. He appeared to be singlehandedly attempting to tackle all of planetary astronomy's desiderata. Looking at the facilities, programs, and staff of the LPL in retrospect, one can see a promising, productive institution, in which many of planetary science's most energetic, capable, and productive workers first appear as graduate students and research assistants. But during the first years of the institution, many of these workers were unknown. The senior members of the staff, with the exception of Kuiper, were all jointly appointed with other departments at the university. Moreover, as the comments quoted above show, there was a perception at headquarters that perhaps Kuiper was spreading himself too thin. There was a genuine concern evident that as Kuiper's various projects grew, his valuable time was spent more and more on administration, and his scientific productivity suffered. Although telesccope domes sprouted like mushrooms on the Arizona mountains after Kuiper's arrival, the largest NASA planetary telescopes would go to other institutions.[105]

With the establishment of NASA and a planetary exploration program, however minor compared with the manned earth-orbital and Apollo lunar landing programs, it was not long before various parties discovered the gaps in the knowledge of the solar system, particularly the inner planets. By the early 1960s, various proponents of planetary exploration, including scientists, space planners, and space engineers, had become convinced

FIG. 2.3. The 61-inch reflecting telescope at the University of Arizona's Lunar and Planetary Laboratory was the first major planetary telescope funded by NASA. Photograph courtesy of NASA.

that most astronomers were not likely to reorient their research interests toward planetary astronomy. In any case, it was becoming increasingly apparent that there was a shortage of all kinds of scientific expertise relative to the recent burgeoning demand. The question to be settled was, who would fund planetary astronomy and what was the best way to go about stimulating the neglected speciality? While various interested individuals and groups urged and attempted immediate action, NASA headquarters officials mulled over the situation, pondering and investigating various solutions. They had the leisure to do so, since planetary exploration was a relatively minor program in the early years, and the possible contributions of ground-based studies, while many, were not clearly defined. Proponents seemed to want to do everything at NASA's expense. Of course intense ground-based scrutiny of the planets would help in the planning and execution of the probe missions. The question was, would its contribution be worth the investment?

THREE

Funds and a Focus (1963–1970)

While many individuals and groups, scientists, engineers, and administrators among them, had loudly proclaimed to NASA the need for stimulating ground-based planetary astronomy, the agency seemed to move slowly on these recommendations. In part this was due to the relatively low priority of the planetary program within NASA, the confidence that probes would get to the planets before the ground-based techniques would pay off, and uncertainty about the propriety of the space agency involving itself in ground-based astronomy, an area that was the province of other agencies. The rise of planetary exploration as an important program and dramatic new ground-based findings on the atmosphere of Mars, which threatened a major element of that program, helped push headquarters officials toward greater involvement in ground-based astronomy. The planetary program provided both funds and a focus for NASA's involvement in ground-based planetary astronomy.

AD HOC SOLUTION: SUPPORTING RESEARCH AND TECHNOLOGY

At NASA headquarters there was some indication that Newell and others were getting ready to take action. The Supporting Research and Technology budget of the Lunar and Planetary Programs office had been used to fund various kinds of planetary observation and analysis since very early in the program. Appendix 5 shows grants and research contracts let to university investigators from 1958 to 1963 in planetary astronomy. Projects supported during the first year or so are weighted toward basic lunar astronomy and celestial mechanics. Other lunar astronomy was being carried out under partial Air Force and Army support, with NASA subsidizing the military funding through fund transfers. More planetary projects and more diversified techniques appear after 1960, and most of the astronomers already engaged in some sort of planetary astronomy were receiving support from NASA, either as Principal Investigators or indirectly by being employed on the PI's grant or contract.[1]

In the fall of 1961, Newell's space science operation was moved out from under Spaceflight Development and elevated to the status of a program office. This gave space science a higher prominence in the agency, and gave Newell, from his new position in the organizational structure of NASA, greater power to move toward a broader-based space science program.[2]

In response to a letter from Donald F. Hornig, chairman of the President's Science Advisory Committee's Space Science Panel, which referred to a "crisis of confidence between NASA and members of the scientific community who participate in the NASA program," NASA asked the Space Science Board to conduct a summer study in 1962 to review the entire NASA space science program.[3] The 1962 Summer Study did not consider the question of expanded ground-based planetary astronomy at any length because "most specialists in this field were not able to participate." The study did recommend, however, that NASA convene a panel to study the issue of required ground-based planetary work, and suggested that John Hall of Lowell Observatory chair the panel.[4] The study addressed the question of the Cloudcroft Air Force lunar and planetary observatory, and whether NASA should operate it after the Air Force established it. The study recommended that, although such an observatory had value, "finding an eminently qualified man to plan and direct it is essential." The study found the Otero County site "ill-advised," and recommended trying to achieve a better distribution in latitude and longitude. It encouraged the establishment of "one or more lunar and planetary observatories for more general and basic research."[5]

The importance of ground-based astronomical research, planetary and other types, was mentioned in a number of places in the reports and recommendations. A year later, a "NASA Reaction to Space Science Summer Study" noted that "work is in progress on ground-based studies of the planets."[6] Earlier, in January of that year, Newell had received correspondence from the SSB reminding him that the issue of planetary astronomy had still not been resolved, and that the committee recommended by the summer study had not yet been formed.[7]

But in the reorganized Lunar and Planetary Programs office, where Oran Nicks had become director, steps were being taken toward paying more attention to planetary astronomy. Nicks, while working at the Missile Division of North American Aviation in 1957–58, had been part of corporate study efforts to define lunar and Mars missions. Through his contact with Urey, Kuiper, and others on the Ranger and Surveyor projects, as well as the subcommittees of the Space Sciences Steering Committee, Nicks had become aware of the potential for ground-based planetary astronomy. In May 1962, Nicks assigned Roger C. Moore of the Lunar and Planetary Programs office "disciplinary responsibility for ground and balloon-based astronomy."[8] By this time, the planetary program had begun to pick up, although in the wake of President Kennedy's Apollo decision the previous year, lunar programs, especially Ranger and Surveyor, had top priority. The first interplanetary spacecraft, Mariner II, was launched in August 1962 and was headed for an encounter with Venus in December. Moore and others prepared letters to observatory directors around the world informing them of the time of Mariner II's encounter, the craft's instruments, and measurements expected to be made with the spacecraft's infrared and

FIG. 3.1. Oran W. Nicks. Photograph courtesy of NASA.

microwave radiometers. In some cases, the letters requested that the observatories conduct various ground-based observations to support the probe's encounter and for comparison with the data returned by the probe.[9]

A PLANETARY ASTRONOMER AT HEADQUARTERS

Nicks recalled that they had been "looking for a professional [astronomer] for quite a while" to coordinate the ground-based planetary astronomy support in the Lunar and Planetary Programs office. In July 1963 Ron Schorn was detailed from the JPL group to headquarters, and shortly thereafter was listed as Acting Program Chief, Planetary Astronomy. According to Schorn, three things were required in order to establish the

NASA planetary astronomy program: (a) a director for the program with a "union card" or Ph.D. in astronomy, in order to have some acceptance within the astronomical community; (b) proof that ground-based planetary astronomy could "affect millions of dollars in the planetary flight program," to impress top NASA management; and (c) an established astronomy department willing to make a long-term commitment to planetary astronomy. Schorn fulfilled the first condition as an advocate in a way that Roger Moore and Ray Newburn had been unable to do. Proof that planetary astronomy could affect the distribution of millions of dollars in the flight program was not long in coming.[10]

Shortly after Schorn's arrival, news came to NASA that JPL's Lew D. Kaplan and Hyron Spinrad, working in collaboration with Guido Munch of Caltech, had redetermined the Martian surface pressure, arriving at a figure about one third that of previous estimates, estimates on which proposed landing spacecraft had been designed.[11] This new estimate had repercussions throughout the astronomical as well as the spaceflight community.[12]

Although their results were not published until January 1964, word of the finding reached Nicks by early August 1963 via Lunar and Planetary Programs science chief Urner Liddel. Liddel attended an astronomy sub-committee meeting at the NASA Ames Research Center on August 9, where Spinrad reported the findings. Concerned about the effect that such a dramatic change in expected conditions on Mars might have on the planetary program's Voyager Mars lander designs, Liddel "urged every astronomer who will be looking at Mars to do what he can to measure its atmosphere. He promised to do all he could with respect to supporting such research, including ground based observations." Liddel asked for suggestions of astronomers to invite to a meeting to discuss the Voyager Mars problem, and "reiterated his plea for an all out attack on the Martian atmosphere. . . ."[13] Nicks told Newell, who then called a conference of "a dozen experts in planetary astronomy" at NASA headquarters on October 1, but found "no agreement among these experts as to the best value."[14] The next opposition of Mars would not occur until March 1965, but Newell and the Space Science office at NASA started looking for investigators to observe Mars starting in the summer of 1964 when the planet would be well placed for the observations. However, there were only "four experienced observatories equipped to undertake these observations," and observing time for the measurements had to be procured "by dint of intense persuasive effort."[15] Several planned Mars missions were cancelled and others were deferred until the Martian surface pressure could be determined with confidence.[16] Since the creation of the planetary program in 1959, when NASA promised that launches aimed at Venus and Mars would be attempted every time that planetary alignments permitted, three-fourths of the "launch windows" had been missed, and now the program faced an uncertain future.

To be fair, the ever-shifting estimates of launch vehicle availability,

FIG. 3.2. Urner Liddel. Photograph courtesy
of NASA.

distressing failures in the Ranger lunar program, the high Apollo lunar
priority, and budgetary problems all played major roles in the lunar and
planetary program difficulties. But that single spectrogram of Mars, taken
with a fifty-year-old telescope using techniques not very different from
those used in the 1920s and 1930s, certainly helped shake the confidence
in accepted planetary data which the engineers had been using as a basis
for design. Ron Schorn recalled, "It blew the lid off everything."[17] In addition, the logical progression of exploratory spacecraft, which would super-
sede ground-based results by returning *in situ* data in time to be used
in planning the next step in spacecraft complexity was now gone, leaving
only the rudimentary Mariner IV and ground-based techniques to pave
the way for the advanced and complex Voyager Mars lander.[18]

The events surrounding the revised surface pressure calculation brought
the planetary program engineers and managers into increasing and closer
contact with the planetary astronomers. It also gave managers suitable
leverage to use in prying loose funds for ground-based planetary astronomy,
since it furnished a graphic and forceful proof of the value of the older

methods to the design and execution of the newer methods. Perhaps if a steady stream of interplanetary spacecraft had been sent to Mars and Venus as originally planned, the *in situ* measurements would have superseded and even discredited ground-based studies. But the admonitions of Kuiper, Newburn, and others who had alerted NASA planners to the problem years earlier now gained an urgency lacking in the more optimistic earlier years.

The planetary astronomy advocates had been seeking for years to establish the critical importance of the field for the planetary exploration program. In the vocabulary of the systems approach to social studies of science and technology, a reverse salient had developed in plans for the most important planetary program in NASA. The atmosphere of Mars had become the critical problem to be solved. Now the planetary astronomy advocates could ally themselves with the planetary program planners over a clear and specific issue.[19]

THE CRITICAL NEED FOR TELESCOPES—MAJOR FACILITIES CONSTRUCTION

Before being detailed to headquarters Ron Schorn had trained in radio astronomy at the University of Illinois, where he had done some work in centimeter- and meter-wavelength lunar radar. In 1962 Newburn had hired him for JPL's growing optical astronomy group, which at that time included Hyron Spinrad. Outspoken and energetic, Schorn now brought the JPL planetary astronomy zeal to headquarters at just the right time. With the Mars pressure redetermination as proof, an assignment from Nicks and Liddel to accelerate the planetary astronomy activities, and a box on the organization chart devoted to planetary astronomy, Schorn and Roger Moore raised "several million dollars" from various lunar and planetary program funds at headquarters. They also tried, but unsuccessfully, to get the Cloudcroft money transferred from the Air Force to NASA.[20]

But raising the funds was only part of the requirement. NASA policy of support to universities and other institutions had evolved by then into an entire University Program, in which the agency would provide funds for research with the aim, not only of buying the research that NASA needed, but of strengthening the recipient and the overall scientific and engineering capability of the nation as well. This grew out of a conviction and a vision of Administrator Webb, and was a policy forged amid the various criticisms that NASA was draining the scientific and technical pool. It also evolved from more pragmatic estimations of what kinds of funding NASA could get through the Bureau of the Budget and Congress. In practice this policy meant that for NASA to fund a major piece of research equipment at a single institution, that institution had to match the gift by contributing funds raised elsewhere to the project, and by making a commitment to use the facilities to enrich the scientific and engineering resources of the country. In short, for a planetary telescope to be built at an institution, that institution had to contribute the pier, building, dome, and other "bricks and mortar" construction, and show in other ways that it was committed to planetary astronomy and teaching. In this way, NASA would get what it needed, the institution would benefit, and the overall scientific and technical pool of talent in the nation would be enriched.[21]

This was the third requirement mentioned above by Schorn: an institution willing to make a commitment to planetary astronomy. Schorn and Moore found two major institutions which looked like they would be both willing and appropriate. First, Harlan Smith, a planetary radio astronomer from Yale, had recently assumed directorship of the McDonald Observatory at the University of Texas and was seeking funds to renovate the 82-inch Struve reflector, and to add a copy of the Kitt Peak 84-inch. Second, Caltech seemed interested in obtaining a 60-inch reflector. The latter would neatly solve the problem of the JPL astronomers' access to a large telescope with a guarantee of planetary observing time, fulfill the University Program requirements, and smooth strained relationships between the JPL group and Mount Wilson, all without NASA having to construct a facility of its own.[22]

Moore and Schorn took their proposals to Newell, along with the suggestion that in addition to funding the basic instrument, NASA should also fund the planetary portion of the operation of the telescope for a certain number of years, so as to guarantee access. Newell approved this, in spite of opposition from within the Lunar and Planetary Programs office, where it was thought the money would be better spent on additional probes. According to Schorn, Newell said he would be willing to cancel a mission for the ground-based program if required. Newell had overcome his earlier skepticism, and was now convinced of the value of ground-based planetary astronomy, and of the necessity of NASA taking action to assure its being carried out.[23]

Caltech and Texas submitted their proposals in March 1964. The Texas proposal proceeded fairly smoothly through the review process. The Caltech proposal, however, developed a serious snag. Caltech had been under the impression that this would be a complete NASA facility, much like JPL, "designed and constructed under the direction of CalTech people, and operated by CalTech for NASA." Since under the sustaining university program only one "bricks and mortar" facility could be given to each institution, Caltech did not want to use up their facility option on the planetary telescope.[24] Negotiations continued into the fall, and in October, after the Texas contract had been signed, Newell reported that negotiations with Caltech were "at a standstill."[25] In November, Newell wrote to Caltech President Lee Dubridge breaking off negotiations.

In his letter to Dubridge concerning the planetary telescope, Newell articulated elements of the NASA policy of support to ground-based astronomy which were to remain in effect throughout the sixties and seventies. First, it was agreed at NASA that the agency would have to take extraordinary steps to secure the lunar and planetary data it needed for the flight programs, since existing astronomical observatories were "almost fully committed to stellar and solar work, and are not available for increasing work in planetary astronomy." That is, NASA had a need for ground-based astronomy which was not being met by existing facilities. Second, "NASA does not deem it appropriate to construct and operate such a planetary observing facility as a NASA facility. It is NASA policy to seek and solicit the interest and support of the university and astronomical community in meeting the NASA program need for additional ground based planetary observations and research, and not to set up activities that compete with the university community for scientific manpower." Hence the requirement that the recipient institution make a "substantial commitment" to the facility.[26]

In part this policy derived from the ground rules of the NASA University Program and pragmatic considerations about what Congress and the Bureau of Budget would permit. But another factor was the feeling that this need for ground-based planetary astronomy was a temporary situation, and that an extensive program of probes would rapidly make such techniques obsolete. This attitude also contributed to the NASA approach of seeking out locations with established astronomical expertise. In explaining why NASA had decided against locating a large planetary telescope at one small institution, Newell explained to Associate Administrator Robert C. Seamans that "a minimum of several years would be required to build up an appropriate staff and construct these facilities, by which time our critical program need for planetary data would be past."[27]

Another factor operating at this time was the overall state of ground-based astronomy. In addition to the demands that the NASA program was placing on astronomy directly, public enthusiasm for space activities manifested itself in the schools and universities. Astronomy departments and curricula, being the most visibly space-related, received a good deal of attention. The August 1964 Whitford Report on astronomical facilities, noting the high demand for graduate astronomers, added:

> Whereas 15 years ago the few new astronomers produced each year were sufficient to satisfy the immediate needs of that era, today there are not enough astronomers either to satisfy the demands of the space program or to keep pace with expanding university requirements. . . . Graduate schools are now flooded with applications in astronomy; enrollment is higher than it ever has been and is increasing at the unprecedented growth rate of 19 per cent a year.[28]

The Whitford Report had made only passing mention of planetary astronomy. After noting that "there are not enough astronomers either to satisfy the demands of the space program or to keep pace with expanding university requirements" the report called particular attention to the demands being felt for planetary observations:

The recent resurgence of interest in planetary astronomy, encouraged by the space program, has created new demands on large telescopes which likewise cannot be met. Commitments to programs already in progress have made it difficult for observatories with large telescopes to divert time to ground-based re-evaluation of many parameters of the planets and their atmospheres, which are of vital importance in planning vehicular missions to points in the solar system.[29]

Unwilling to commit itself to a long-term facility for what it considered a temporary need, and mindful of criticisms that it was draining the nation's pool of scientific personnel and creating additional pressures on disciplines it wished to have as allies, NASA had arrived at a policy which would allow it to solve several problems at once. Reviewing the Whitford Report in *Science,* John Walsh wondered, "since many disciplines are eyeing the space treasury, and since NASA claims it is close to being financially overdrawn on its commitment to land a man on the moon in this decade, it is difficult to see why the space agency would clutch at a chance to become a major financier for a field that it has heretofore managed virtually to ignore."[30] The short answer to Walsh's question is that NASA needed to know the atmospheric pressure on Mars.

NASA sought to alleviate the shortage of telescope time for planetary observations by building telescopes and other instrumentation at locations where there already existed strong astronomy programs. By including language in the contract requiring a certain percentage (often 25–50%) of the telescope's time be spent on research needed by NASA, the agency could assure that interested investigators would have the observing time and facilities they required. By locating the instruments at institutions with competence in astronomy, nonplanetary astronomers' opportunities would be enriched and, hopefully, better relations would be established.[31]

In January 1965, Donald Hornig, President Johnson's science advisor, asked NSF to convene an ad hoc panel representing all federal agencies involved in funding astronomy to assess the Whitford Report and coordinate their various plans for alleviating the problems of ground-based astronomy.[32] As part of the NASA contribution various managers from the astronomy, planetary, and solar portions of the Office of Space Sciences reviewed in light of the Whitford Report's recommendations the more or less ad hoc policy NASA had been following. They endorsed the Whitford Report as a conservative statement of the needs of astronomy. They thought that NSF should remain the agency with primary responsibility for federal funding of astronomy, but consistent with NASA's program needs NASA would continue, and slightly expand, its support of certain areas. These and other assessments were put into a formal letter from Hugh Dryden to Leland Haworth, which was transmitted with an attached copy of the staff paper and a lengthy table of current and proposed instruments being supported by NASA. All of this was then incorporated into the Interagency Committee's final report to the White House.[33]

NASA was at this time moving toward planetary exploration as its major post-Apollo goal, and indeed would actually cut back on the stellar and galactic space astronomy programs that partially justified its funding of ground-based astronomy. NASA was keeping the planetary motivations for ground-based astronomy very low key. The need for the large telescopes mentioned by Dryden arose from requirements for planetary spectroscopy, particularly of the atmosphere of Mars, yet no mention is made of this in the NSF papers. It was, after all, the planetary flight program offices which had contributed the funding for the telescopes. Inferences may be drawn from this concerning the status of planetary astronomy within the overall astronomical community and the NSF, as well as the problem of encroaching on another's turf involved in NASA supporting ground-based astronomy. Newell later recalled, "At the time I was with NASA we seized every opportunity to make it clear that we were not trying to encroach on the rights and responsibilities of NSF, even though NASA had large sums of money slated for astronomy."[34]

The Interagency Committee served to clear the air and make explicit a number of policy matters which had heretofore been handled more or less informally. A coordinated plan emerged to implement some of the desiderata of the Whitford Report, NASA's mission needs were explicitly recognized and circumscribed, and NSF emerged with the blessing of the National Science Board and the White House as the formally designated "lead agency" for ground-based astronomy.[35]

The decision by NASA to establish its instruments at locations of recognized astronomical expertise and planetary interest also had its pragmatic motivations. NASA could not afford to wait for the new generation of planetary astronomers to be produced. It needed to get the new telescopes constructed and into operation quickly, and needed to have them located where established experts could rapidly make use of them. When the Texas contract was signed in September 1964, plans were to have it in operation by the April 1967 opposition of Mars. NASA had missed out on a number of opportunities to explore Mars and Venus, but plans were being laid to make solar system exploration a major effort of the post-Apollo program.

MARS PRESSURE

On January 30, 1964, President Johnson had written to Administrator Webb asking what plans NASA had for the post-Apollo program. In order to reply, Webb established a Future Programs Task Group to make recommendations. In the Lunar and Planetary Programs office, Don P. Hearth chaired an ad hoc Planetary Study Group to prepare a submission to Newell. In April, the group made an interim report outlining the objectives of the planetary program for the 1970s. Scientific objectives, in order of priority, were: biology, geology and geophysics, atmospheres and meteorology. Targets for the planetary program, in order of priority, were: Mars; Venus; Asteroids, Comets, Jupiter, Mercury ("in alphabetical order"); Saturn and outer planets. The top-priority program recommendation was to send Voy-

FIG. 3.3. Don P. Hearth. Photograph courtesy of NASA.

ager orbiter and lander missions to Mars starting with the 1971 opportunity. In support of these missions, the study group recommended that "every attempt be made (on a No. 1 priority basis) to definitize the atmosphere of Mars. using flybys, entry probes, and ground-based observations. An intense effort by astronomers, for this purpose, should be conducted during both the 1965 and 1967 oppositions and other opportunities such as quadratures for these time periods."[36]

In August 1964, while the Whitford Report was being released, the Space Science Board sent Newell a position paper on the post-Apollo space science program. The position paper was an outgrowth of President Johnson's January request, as well as progress in the Apollo program which indicated that the manned lunar landing might be achieved even earlier than previously thought. The position paper first recalled that in March 1961 the SSB had recommended "that scientific exploration of the Moon and planets should be clearly stated as the ultimate objective of the U.S.

space program for the foreseeable future." The new recommendation was even more explicit. The SSB endorsed the use of a visible and understandable goal to "focus attention and energies." While it endorsed a continuation of all types of space research, the exploration of Mars and the other planets, primarily by unmanned spacecraft but leading eventually to manned exploration, was assigned first priority for the 1970s. Mars was chosen because of the possibility of extraterrestrial life, and because it provided the opportunity to explore another planet similar to the earth.[37]

Schorn had finished his detail at headquarters and returned to JPL the preceding July. Roger Moore had left earlier to take a position with the RAND Corporation's Planetary Sciences Department. In June, William E. Brunk from NASA's Lewis Research Center in Cleveland was working at a NASA exhibit at the New York World's Fair, when he received a phone call from Urner Liddel offering him a job at NASA headquarters as Planetary Astronomy Program Chief. At that time, Brunk was unaware that NASA even had a planetary astronomy program, and what little he knew about the planets he had learned in an undergraduate astronomy course. With some graduate astronomy training from Case Institute of Technology, Brunk had worked at Lewis during the 1950s analyzing supersonic airflow and heating. After Sputnik and the transformation of Lewis from a National Advisory Committee on Aeronautics lab to a NASA lab, Brunk suddenly found himself the "lab astronomer" in an engineering organization caught up in the space enthusiasm. He spent a great deal of time giving informal and formal courses in astronomical aspects of spaceflight, and reviewing and translating the papers that began to flow to Lewis from the new NASA headquarters.

He returned to Case to finish his Ph.D. in astronomy, and defended his thesis in the summer of 1962. In the interim he had served as Lewis's representative on Roman's Astronomy Subcommittee since 1960. While Brunk was not the seasoned planetary astronomer NASA had been seeking to direct the program, he was a quick study. He was a longtime member of the NACA/NASA family, accustomed to engineering and experienced in problems encountered in aerodynamic entry. He had excellent professional qualifications as an astronomer in both theoretical and practical aspects of the discipline. Under his direction the planetary astronomy program at NASA would go from an ad hoc temporary solution to an ongoing, integrated part of planetary science. Planetary astronomers would gain sustained access to new, large, state-of-the-art optical, radio, and radar instruments in superior locatons. By 1970, the third and fifth largest telescopes in the United States and a host of intermediate instruments would be custom designed for, and explicitly dedicated to, planetary astronomy under the auspices of NASA.

Brunk arrived in August 1964 to find his new desk "piled high with mail and papers." He inherited from the efforts of Liddel, Moore, Nicks, Roman, Schorn, and others a solid core of supported research, intermediate

FIG. 3.4. William E. Brunk. Photograph courtesy of NASA.

instruments under construction or negotiation, and a number of petitioners for support. Kuiper's 60-inch at the Lunar and Planetary Laboratory was nearing completion, as was Dirk Brouwer's 40-inch astrometric telescope at Yale. The upgrading of the spectrograph and secondary mirrors of the 82-inch Struve telescope at Texas was underway, and negotiations were proceeding toward contract for the projected 84-inch telescope there. The Caltech 60-inch telescope was mired in policy problems that were over Brunk's head. Proposals for several other large- and moderate-size telescopes were in from other groups, including a New Zealand station with two telescopes for the University of Pennsylvania and a joint Northwestern University–New Mexico State 84-inch instrument. The university research being supported totaled over two million dollars, around half the amount spent by NSF on all astronomy research grants, and fully dwarfing the amount spent by NSF on planetary research (see figure 3.5).[38]

Brunk set about getting to know his principal investigators, learning the peculiarities of planetary astronomy, studying aspects of geophysics,

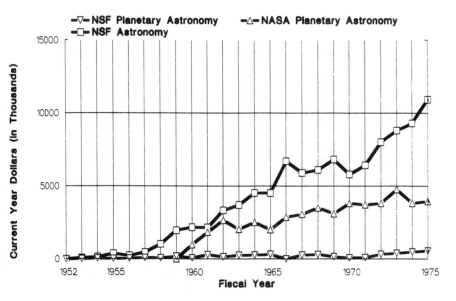

FIG. 3.5. Comparison of NASA and NSF support to ground-based planetary astronomy. (Research grants only. Major telescope construction and NASA in-house research not included.)

and looking at the proposed instruments. He recalled that the needs of the flight program "gave almost complete direction" to the planetary astronomy program. The ground-based lunar research, which had been going on since the earliest days, was high on the agenda but would soon lessen as the Ranger, Surveyor, and Lunar Orbiter spacecraft mapped, scanned, probed, and sampled the lunar surface. At any rate, most of the lunar work was directed toward the engineering requirements of the Apollo program. The most pressing planetary concern was the atmosphere of Mars. The 1964–65 opposition was drawing near; large-telescope time had to be wrangled for high-disperson spectroscopy, and results had to be analyzed as soon as possible and fed to the Voyager engineering staffs. The next most pressing concern was getting the new large planetary telescopes under construction so that they would be ready for the April 1967 opposition of Mars. While their design characteristics would make them valuable instruments for all kinds of astronomy, it was demands for planetary observations and especially spectroscopy of the atmosphere of Mars that drove their special design: large aperture for light gathering, long focal-length for image scale so that small features on the disc could be resolved, coudé focus and excellent spectrographs, and total computer control for following the motions of the planets during long exposures.

Mars would be entering a series of favorable oppositions, and in an almost Lowellian retrospective, observatories and telescopes would spring up especially for it, the planet and its atmosphere would be scrutinized for signs of life, astronomers would argue about the evidence, other astronomers would scoff at what they saw as marginal efforts, an excited and eager public would follow the adventure, and there would be talk of signaling the Red Planet. Yet these oppositions would be quite different from earlier ones. This time, instead of being backed by private fortunes, the new scrutiny of Mars would be done with the funding and urging of the government. This time the world's largest telescopes would be systematically and aggressively used. And this time, communication signals would indeed be sent *and received* from the Red Planet—not to indigenous Martians, but to robot explorers, Mariners and Vikings, sent across the solar system. There was yet another difference, which added both high scientific stakes and human drama to these Mars oppositions. The Soviet Union was also headed for Mars, and uncertainty prevailed about how well or indeed whether they sterilized their spacecraft. Only one spacecraft could win the race; only if that was a U.S. spacecraft could investigators be sure of finding it uncontaminated by terrestrial organisms, organisms which just might find the climate to their liking and forever spoil the chances of a definite answer to the question, "Is Mars inhabited?"[39]

Engineers in Brunk's office, thinking of the problem along lines familiar to them, suggested putting out a request-for-proposal to determine the pressure, density, and composition of the Martian atmosphere. A bidder would then be contracted to perform the work and come up with an answer. More familiar with the ways of astronomers, Brunk recognized this was not the way to approach scientists. He preferred to find astronomers with the required competence and try to interest them in tackling the problem.[40] In this as elsewhere, Brunk's dual familiarity with the cultures of engineering and of astronomy proved extremely useful.

The first priority was to get as many already interested researchers as possible to make observations during the upcoming opposition of Mars. Mars would come to opposition on March 9, 1965, and would be at quadrature (where observations to determine its atmospheric characteristics would be most effective) in December and May. These observations would then serve as a check on the Mariner IV radio occultation experiment in July. Late in the fall, Brunk took a trip to various institutions to meet some of the astronomers in the program, and also to check on plans for the Martian opposition.

At McDonald, where the renovation of the 82-inch spectrograph was underway, refiguring of the mirrors was complete. At Lowell, Brunk found the new planetary data center building under construction. Funded through NASA's facilities grants, the building was to serve as the western depository and research center for planetary patrol photographs. John Hall and the Lowell staff were assembling copies of images from most major observatories, and wanted to establish a uniform system of calibration to be used by all of the groups involved in planetary patrol work (the JPL group,

Clyde Tombaugh and Brad Smith at New Mexico State, the Lowell group, and the eastern planetary data center under Adouin Dollfus at Meudon).[41]

At New Mexico State, Brunk met Tombaugh, Smith, and their group. He was amazed at the quality of composite photographs they were able to make and at the other parts of the program which they carried out with only very modest telescopes, and found them enthusiastic and capable. Later that month Newell himself visited New Mexico State, and in his report praised the work of the group.[42]

At JPL Brunk just missed Frank Gray, who was supervising the transfer to France of a new kind of spectrograph which would be as revolutionary as Kuiper and Cashman's lead sulfide instrument was after the war. Pierre and Janine Connes had visited JPL, where they perfected a fourier-transform interferometer—an infrared spectrometer capable of unprecedented spectral resolution. It was being shipped by the Air Force to be used by the Connes and JPL's Reinhardt Beer on the 76-inch Haute Province instrument for the upcoming Mars opposition. But the JPL optical astronomy group was having problems. They had been reduced in size due to funding problems, leaving Ron Schorn as the only full-time Ph.D. astronomer in the group. Ray Newburn had been transferred to another section of the lab to work on planetary models for mission design. Frank Drake had recently left to take a position at Cornell. Hyron Spinrad was now only on the payroll for a fifth of his time, and R. H. Norton had been transferred to the JPL planetary atmospheres group. Charles Capen was still in residence at Table Mountain, carrying out the photographic planetary patrol with the 16-inch and in close touch with his associates at New Mexico State. For the upcoming Martian opposition Schorn would be at Table Mountain conducting a photographic Mars patrol, using combinations of filters and emulsions duplicating the spectral response of the Mariner IV television cameras. Schorn and Spinrad, with Harlan Smith, would be using the 82-inch at Texas during the quadrature times for photographic spectroscopy as far into the red as the equipment would allow. Robert Younkin would have time on the Mount Wilson 60-inch during March for photoelectric spectrophotometry, and Thomas Bull would have access to the McMath solar telescope at Kitt Peak for infrared interferometry.

At Caltech Guido Munch and Gerry Neugebauer would be using the Mount Wilson 100-inch for spectroscopy, and Munch planned broad-band infrared spectroscopy on the 60-inch. Bruce Murray had some twilight time on the 200-inch for high-spatial resolution infrared photometry.

At the Lunar and Planetary Laboratory, Kuiper was hoping to have the new 60-inch in operation by the end of March for infrared spectroscopy and ultraviolet photography to study the "blue clearing" phenomenon. Using the Kitt Peak 84-inch, Kuiper planned photographic and infrared spectroscopy.

In the course of the regularly supported Harvard multicolor photoelectric photometry project, Donald Menzel was planning observations from the 16-inch at Boyden Observatory, South Africa, and La Houga, France. John Strong was planning another balloon flight to about 80,000 feet for infrared spectrometry.[43]

In all, Brunk found that a large and varied program for studying Mars had been assembled in a remarkably short time. In large part this was due to the fact that an informal network of planetary astronomers had been in operation, in one form or another, for more than five years.

The Martian opposition and quadratures came, and Mariner IV closed in on Mars. It carried the usual complement of particles and fields detectors, plus two additional experiments. The spacecraft's S-band radio would be used to measure the density of the Martian atmosphere as the craft was occulted by the planet during the flyby, and a television system was to take the first high-resolution pictures of Mars during the encounter. This was not without hazard. The occultation required relinquishing contact with the spacecraft after the pictures had been taken but before they could be played back to earth, in the faith that contact could be re-established after occultation. But the atmospheric pressure was a vital datum to NASA planners, and worth the risk. The television experiment had been designed by Robert B. Leighton, Murray, and other Caltech and JPL colleagues. Leighton had pioneered a device for taking photographs of Mars and other planets with the 200-inch at Palomar, using moving picture film and automatic compensation for fluctuations in seeing. Although the television system was "state-of-the-art" for the time, it could only return a series of about twenty-one pictures, covering about one percent of the Martian surface, as the craft sped by. The best resolution would be comparable to earth-based telescopic photos of the moon.

Leading up to Mariner's encounter with Mars, the NASA Voyager planning activity, a Space Science Board Summer Study on Exobiology, and a spate of articles in the scientific and popular press raised hopes and expectations for this first close-up view of the Red Planet. And indeed, the Mariner IV mission was a technological triumph, and produced startling scientific results. Ironically, however, its very success dealt a sharp blow to the planetary program and with it the fledgling planetary astronomy program.[44]

The radio occultation experiment, combined with the results of the intensive ground-based observations, produced estimates for Martian surface pressure which converged around 10 millibars, even lower than previously thought. John Naugle's Voyager Capsule Advisory Group was satisfied that the mutually reinforcing ground- and probe-based measurements eliminated the need to send an atmospheric entry capsule on a mission in 1969, and that mission was eliminated.[45] However, the lower surface pressure cast doubt on whether the Voyager lander could be designed to survive landing and also stay within the weight limitatons of the Saturn IB-Centaur combination launch vehicle. Newell postponed the request for proposals for Voyager landing capsules.[46]

Potentially more damaging to the overall program than the conclusions about Mars' atmosphere were the television pictures of the planet's surface. Nothing at all resembling canals was seen. In fact, no earthlike features at all could be glimpsed. While craters had been anticipated by various observers, no one was prepared for the moonlike appearance of Mars. Moreover, the heavily cratered areas seemed to indicate that no change had taken place there in perhaps two to five billion years. The fact that the cratered terrain was so well preserved led the TV team to conclude that "no atmosphere signficantly denser than the present very thin one had characterized the planet since that surface was formed," and the absence of any erosion features led them to believe that no significant quantities of liquid water could ever have existed there. The absence of a magnetic field and the absence of any kinds of stressed features on the surface argued for the planet being "dead" internally as well.[47] While a few pictures of less than one percent of the Martian surface, at fairly crude resolution, could not proclaim the planet truly dead, what was seen in the pictures was so shocking in view of the speculations concerning Mars that it proved a severe blow to the proponents of further Mars exploration. NASA and the SSB had hung the entire solar system exploration program on the urgency of the Mars project, and that urgency rested largely on the similarity of Mars to the earth and the likelihood of finding life there. With both of these premises seriously in doubt, opponents had gained considerable leverage. The Mars program (which continued to become more expensive and more complex each day) and the space program in general now faced growing criticism.

Nonetheless, Voyager planning continued. The Voyager Science Strategy Group met in late September and agreed to establish a Voyager "supporting research and technology task" to support ground-, balloon-, and rocket-based astronomical observations for the 1966 and 1969 oppositions. Naugle met with Brunk, Nancy Roman, and Urner Liddel, and agreed that Brunk would check on the further use of ground-based telescopes and Roman would investigate the use of the orbiting astronomical observatory for Mars observations.[48]

In October, partly due to the realization that the thin Mars atmosphere required a heavier spacecraft (to carry larger parachutes and retrorockets), the Saturn IB-Centaur development was cancelled and plans were announced to launch two Voyagers on a single Saturn V. In December, with cuts in the NASA budget and the Surveyor and Apollo projects consuming more funds, Voyager was deferred until the 1973 opportunity, and a pair of advanced Mariner flybys scheduled for the 1969 opportunity. Discovering a "hitherto well-concealed interest in Venus," NASA ordered JPL to make some minor modifications in a spare Mariner spacecraft and send it to Venus in 1967.[49] The critical problem of the atmosphere of Mars had demonstrated the synergism of ground- and space-based techniques in studying the planets. The operation was a success, but the patient might still die since the results upset a program based on questionable planetary data and influenced by broader social and political expectations.

While Mariner IV was approaching Mars, the Space Science Board was conducting another summer study, at NASA's request. The Working Group on Planetary and Lunar Exploration paid particular attention to ground-based studies as part of an overall research strategy. They noted that certain planetary observation programs

> have not had the priority they should have been accorded if they had been considered in the light of their importance to an expensive national program. Priorities in astrophysical [sic] observatories are, quite rightly, awarded on the basis of over-all contribution to astronomy. In order to introduce a different set of priorities it will be necessary for NASA to continue, and, probably, expand its support of ground observations in a few locations.[50]

The group endorsed the two NASA-sponsored telescopes already under construction (in Texas and Hawaii; see below), but warned that they might be insufficient. It also noted that the Whitford committee's recommendations would not meet the additional needs since the committee "had no information on the extent of the large scientific program [for solar system exploration] planned by this study group." They also were concerned about the geographical distribution of facilities and the absence of a Southern Hemisphere instrument. The group added that "it is dangerous to leave the demands of a large national program subject to the priorities of groups which are not directly involved," and stated that "NASA is not being asked to support academic astronomy but to support its own interests." The group concluded with two formal recommendations: (1) that NASA give "very high priority to the construction of ground-based telescopic equipment to the extent required to provide maximum support to the planetary flight program"; and (2) that, as the previous 1962 study had also recommended, a study of the required ground-based planetary astronomy be conducted, "with the aim of identifying the gap between existing and projected instruments and the needs of planetary exploration."[51]

At its November 1965 meeting the SSB accepted the Working Group's recommendation for a study of needed ground-based planetary astronomy, and decided to bring it to the attention of the National Academy of Sciences Committee on Science and Public Policy (COSPUP), the sponsor of the Whitford Report.[52] Richard M. Goody and Herbert Friedman took the recommendation to COSPUP on January 16, 1966, but COSPUP "was reluctant to undertake further study of this topic at this time, feeling that all that was necessary was a pulling together of existing information."[53] The reception that Goody and Friedman got indicated that, in spite of NASA's realization that stimulation of the field was required, and in spite of the agency's efforts to support ground-based planetary astronomy, there still had not been a great change in the estimation of planetary research among traditional astronomers. The planetary astronomy enthusiasts had succeeded, after many years of persuasion and maneuvering, in convincing NASA to put a small amount of its resources into their field to support its own interests in planetary exploration. Support for ground-based activities sufficient to a vastly expanded planetary exploration program would require allies in greater numbers and of greater influence.

John Hall had chaired a conference on the Martian atmosphere for NASA in October 1966, and he reported to the SSB that "several spectroscopists left the impression that it might be difficult to obtain enough observing time at telescopes of sufficiently large aperture to carry out their programs." He reported that Harry Hess had written to the directors of a number of observatories on behalf of planetary studies, but that the four replies received, and the lack of response from the rest, "indicate that the Martian programs have no priority among others which bear little or no relation to the space effort." Hall saw a "serious lack of ground-based telescopes at which timely planetary programs can be assigned top priority."[54]

Hall called attention to the excellent conditions in Chile, and the fact that the Association of Universities for Research in Astronomy Chile site was at a latitude where Mars would be almost directly overhead during its most favorable oppositions, and two or three hours of longitude east of the southwestern American observatories. "It would seem very desirable . . . that a third large planetary telescope be situated in Chile."[55]

Hall also spoke to the question of graduate training. Of the twenty spectroscopists at the conference, equally divided between laboratory and observational workers, only two had done most of their work on planetary topics, and in his judgment their participation stemmed from "loyal support of a major government program and, in some cases, the financial support this program has given them. In a sense, *NASA has had no other recourse than to resort to a large measure of financial suasion*" (emphasis mine). While astronomy graduate students are exposed to many courses in stellar and galactic structure, Hall noted, few ever get the chance to take a graduate level course in planetary studies. "Astronomers are very likely to be trained in graduate schools at which planetary work is either ignored or low caste." Is it not time, Hall asked, for geology, geophysics, and geochemistry departments to start exposing their graduate students to the use of telescopes?[56] Hall assembled an ad hoc panel, which met at work sessions during 1967. Four-and-a-half years after the first recommendation at the 1962 Iowa City Summer Study (which had suggested Hall as chairman), a thorough assessment of the state, status, and desiderata of planetary astronomy was being made.[57]

The conference on planetary spectroscopy which Hall had attended was part of an overall strategy which Brunk had arrived at. Given the perceived lack of interest of the rest of the community in planetary work, Brunk had wondered how to get someone who was a competent worker in spectroscopy interested in applying the technique to the planets. What he discovered was basically the cogency of L. A. Manning's remarks to Homer Newell after the 1960 SSB conference on the atmospheres of Mars and Venus: once you get a scientist hooked on a problem, they will automatically want to pursue it. Brunk discovered that by introducing spectroscopists and other workers to a problem, many would get hooked on it. Newell had not found this convincing in 1960, and even less so had Newell found

Manning's argument that NASA had a responsibility to support ground-based planetary astronomy (see chapter 3). But the events surrounding the Voyager project, as well as Newell's many discussions with astronomers, had helped change his mind.[58] This is yet another instance where arguments that NASA officials had once found unconvincing took on new significance in light of the Mars atmosphere problem.

Indeed, Hall had been most impressed by the new techniques that planetary workers had started to apply to Mars. While it is true that, in order for planetary astronomy to be prosecuted more vigorously money had to be injected, equally important was the need for a focus. The early suggestions by JPL and others had ranged over all possible studies, had emphasized synoptic patrol-type work, and consequently seemed rather diffuse. This was what Roman had found so objectionable, and what most went against the grain of the system of scientific funding—the "NSF mode" in which one submits a focused, specific proposal demonstrating a problem to be attacked, the importance of the problem, the methods to be used, and the prospects for success. It was the Voyager Mars program that provided both: funds and a focus.

THE MCDONALD 107-INCH

With Harlan Smith's move from Yale in 1963 to assume directorship of the McDonald Observatory in Texas, the NASA ground-based planetary program found one of its vital desiderata: an institution willing to make a substantial commitment to planetary astronomy, and also willing to contribute the cost of everything but the primary instrument itself. Mars expert Gerard DeVaucouleurs had been there since 1960, and the astronomy department and program was expanding. In addition, the characteristics of the site were well known, the latitude was desirable for planetary observations, and the isolation of Mount Locke seemed to guard the site from urban encroachment. The Texas telescope was from the start designed for planetary astronomy, particularly spectroscopy. The original plans in early 1963 called for using the design of the Kitt Peak 84-inch, since the existing detailed plans could be used to get the instrument into operation by the spring 1967 opposition of Mars. The contract was signed on September 22, 1964. But litigation concerning the Kitt Peak instrument made release of the detailed plans problematic, so Texas contracted Charles W. Jones Engineering of Los Angeles to design the telescope. But the tight schedule for completion argued against pouring and finishing such a large blank for the primary mirror in time. A 105-inch blank was available, however, and with the press of time, designers took the 105-inch figure as their starting point. Meanwhile, Jean Texereau, chief optician of the Paris Observatory who was at that time refiguring the Struve telescope's coudé optics, found fatal flaws in the mirror blank. They now had detailed engineering plans for a 105-inch telescope, but no blank for the primary mirror.[59] The

Corning Quartz Factory, however, had developed a process by which a number of slabs of silica could be sagged together to make a large, stress-free blank. By the end of 1965, a primary blank of 107-inches had emerged from the Corning furnace and was ready for figuring.[60] Hence the most important technical parameter of the telescope—its light-gathering power expressed by its clear aperture—was far from the simple deduction of some scientific desideratum. Rather, it was the compound product of a variety of factors and circumstances.

Meanwhile Smith and the University of Texas had gotten partial NSF support for the piers, building, dome, roads, aluminizing chamber, mirror-handling crane, electronics, and finder telescopes.[61] By the spring of 1966, with the mirror size fixed, and the surface being figured, fabrication of the telescope tube and other mechanical parts could begin. With the spring 1967 Martian opposition now unattainable, the target date for completion of the telescope was set at October 1967. Westinghouse's electric boat division was awarded the subcontract. "During contract negotiations the project was assured high priority; during construction large military and commercial jobs took precedence over the telescope," Smith noted in his final report, and final assembly of the mechanical portions of the telescope was accomplished a year late and about a million dollars over cost.[62]

With the May 1969 opposition barely six months away, the telescope was installed under Westinghouse direction during the fall of 1968. Although the fabricated mechanical parts were "excellent," errors during installation led to severe damage of the large worm gears during testing, and the "wiring, drive system, and controls proved basically inadequate as well as full of errors." Texas took over from the Westinghouse team, and began repairing the damage to the drive gears, rewiring the controls, and otherwise trying to get the instrument into service. Although still months away from operation, the telescope was dedicated on November 26, 1968. Perhaps characteristic of the history of the instrument, a freak Thanksgiving Day blizzard soaked the Marfa, Texas airport, and a chartered plane containing NASA headquarters officials among others on their way back to Washington came to an abrupt halt during takeoff as the plane's wheels sank in the soft mud.[63]

With a patched control system, the NASA–McDonald 107-inch telescope saw first light on the very night of Martian opposition, March 9, 1969, as Mariners 6 and 7 sped toward a July flyby of Mars. A Connes-type interferometer built by JPL's Reinhard Beer began engineering tests that night, and a week later Goddard's Rudi Hanel was testing a second interferometer in the coudé room. Two years and two Martian oppositions late, and almost two-and-a-half times the originally estimated $2.1 million NASA cost, the McDonald telescope was in operation.[64] Planetary astronomers now had guaranteed access to the world's second largest optical telescope.

It was a pleased and relieved Bill Brunk who, five years after arriving at NASA headquarters to find the McDonald telescope under negotiation, addressed the gathering at the "scientific dedication" of the telescope in October 1969: "I must admit that there were several times when I wondered whether I would actually stand under this telescope."[65]

Less than a year after first light, however, the instrument was intentionally damaged by a new night assistant. On February 5–6, 1970, the assistant attacked the primary mirror with a four-pound hammer and then fired seven bullets into the glass from the distance of a few feet. He had been showing signs of stress, and "although psychological interpretations are hazardous, it appears that by lowering the telescope to the floor, walking into the tube and emptying his revolver at point-blank range into his image in the primary mirror, he may have been simultaneously symbolically committing suicide as well as trying to destroy a great instrument which he felt was beginning to dominate him," Smith dryly commented in a report to NASA.[66] Don Davidson, chief optician at the firm which had polished the primary mirror, Johnnie Floyd, University of Texas chief engineer, and the Astronomy Department's Robert Tull drilled out the areas around the bullet holes; the damage reduced the light-gathering power and added some diffraction, both of which were less than one percent, or equivalent to about a month's weathering of the mirror.[67] NASA's first major planetary telescope, the second largest telescope in the world, after a perilous development, was finally in operation.

THE MAUNA KEA, HAWAII, 88-INCH

The breakdown of the Caltech negotiations in late 1964 left Brunk and NASA with funds, but no firm location for the next large planetary telescope. Kuiper had told Roman in late 1962 of the merits of Hawaii as a possible location for a NASA telescope, offering to put a copy of the Lunar and Planetary Laboratory's NASA-funded 60-inch there. Having merely heard of the excellent seeing, Kuiper was soon off to investigate it personally. He made a survey of possible Hawaii sites in 1963, and while investigating Haleakalea, discovered that its otherwise excellent seeing was marred by fog and mist which collected in the extinct volcanic caldera below and occasionally drifted over the summit. "But as Kuiper later told the story, the last thing he saw as the fog closed in was Mauna Kea, standing clear in the sunlight."[68]

In April 1964, Newell reported that, while visiting Hawaii to check on the University of Hawaii's research grants with NASA for studies of the aurora and airglow, he "picked up one disturbing bit of information. It appears that Kuiper, in carrying on his investigation of seeing conditions on Mauna Kea, has very much oversold the idea that an observatory might be built there under the auspices of NASA." Kuiper was supposed to be conducting site testing, but Newell had heard that a road to the summit of Mauna Kea was under consideration.[69] Not only was a road under consideration, but, according to one account, in May Kuiper was sitting next to the driver of a bulldozer as it carved its way from an existing roadhead at 2,800 meters elevation all the way up to 4,155 meters, an operation authorized and paid for by Hawaii Governor Burns. In August, the Hawaii Chamber of Commerce sent Newell "an excellent speech presented by Dr. Gerard P. Kuiper . . . at the dedication ceremony held on July 20, 1964, for the completion of the Mauna Kea summit road and the observatory station, located on the island of Hawaii." Kuiper, Newell was told, "is convinced that we do have a site which could develop into a mecca for scientific research," and had already submitted preliminary proposals to Hawaii officials for development of the site. Kuiper was right about the quality of the Hawaii site, but he would not get the site or the telescope.[70]

In November 1964, with the Caltech negotiations stalled and ready to be terminated, Urner Liddel wrote to the University of Hawaii to "eliminate some of the confusion" which they had expressed "with regard to your potential colleagues in this matter." Liddel assured the university officials that they "must be intimately involved in any future action which is taken by this office [Lunar and Planetary Programs] with respect to the development of another observatory." But, Liddell emphasized, Hawaii would have to have a stateside partner with the required managerial ability, professional staff, and ability to raise funds. Apparently, NASA headquarters had decided that if they could not get the telescope at Caltech, it would go to Hawaii. But Hawaii had no astronomy department or program, and NASA was looking for knowledgeable partners.[71]

Kuiper lobbied headquarters hard, and Newell replied that proposals from the Lunar and Planetary Laboratory would be considered with those from other observatories contacted. Unsolicited proposals came in from Menzel at Harvard, Albert Whitford at Lick, and John T. Jeffries of the newly created Astrophysics and Atmospheric Physics Section of the Hawaii Institute of Geophysics.[72] By early February Newell was writing to George P. Wollard and Abraham Hiatt at Hawaii summarizing a recent meeting. Hawaii it would be for the telescope, but the actual managerial arrangements were not yet set. "NASA interest in strengthening ground-based planetary research in support of our Voyager program might be a useful means for initiating the use of the Mauna Kea site." NASA needed assurances from the University of Hawaii that they would contribute the usual components. Further, an advisory committee should be established to give Hawaii the professional astronomical advice it would need, and a competent director for the observatory was required before the negotiations could go further. Finally, Hawaii was to submit a formal proposal.[73]

By the end of March all signs led toward accepting the Hawaii proposal. "We went out on a limb with Hawaii," Brunk later told a reporter, adding that it was simply the best proposal of the lot. While Jefferies, a theoretical solar astronomer, had no previous experience in developing telescopic sites or constructing telescopes, his proposed copy of the Kitt Peak 84-inch, along with the promise to develop an astronomy department with planetary emphasis, were strong points in Hawaii's favor. In addition, having the managers of the telescope on site rather than on the continent, and the university's commitment to provide all support facilities, helped make the risk seem worthwhile.[74]

When the contract was signed with the University of Hawaii in July, "Kuiper was furious. To NASA and to his colleagues, he maintained that Jefferies was not competent to build a night-time telescope. The University of Hawaii was in way over its head. For years thereafter he told people— strangers—how John Jefferies had stolen Mauna Kea from him."[75]

Indeed, NASA seemed to be violating the very policy it had so frequently enunciated. While Hawaii seemed to be demonstrating a commitment to building up planetary expertise and was willing to contribute the usual facilities, at the time it must have seemed a "long shot." But such was the excellence of the site, the demonstrated enthusiasm and willingness of the proposers, and NASA's misgivings about Kuiper being overextended, that for its third planetary telescope NASA stretched, if not violated, its own policy.

Jefferies moved ahead to build up the staff. From Lick he hired Hans Boesgaard to be chief engineer. From Lowell came William Sinton, drawn no doubt by the promise of the best infrared site in the United States, perhaps in the world. By 1967 Hawaii's Institute for Geophysics had spawned an Institute for Astronomy. By March 1966 the Charles W. Jones Company in Los Angeles had completed conceptual designs and was proceeding with the detailed design of the telescope. Corning was fabricating the primary and secondary mirror blanks, and turned out an unflawed 88-inch clear aperture primary. Site testing had been narrowed to Mauna Kea. Jefferies chose a site barely below the summit, "reserving the summit itself for the 'ultimate-inch' telescope that he hoped might someday be built there."[76]

They were aiming at having the observatory in operation for the May 1969 Martian opposition. Like the Texas telescope, Mauna Kea's observatory would miss its first appointment with Mars. Groundbreaking ceremonies took place in 1967, and work on the telescope progressed. Unlike the Texas telescope, however, the problems in Hawaii were with the site and building. The instrument sat completed in Hilo, while the building contractors battled 100 km/hr winds, winter blizzards, and the oxygen-thin air at the summit of Mauna Kea. In July of 1970, two years late, over budget, and, like Texas, with mechanical and electrical control problems, the Mauna Kea 88-inch was finally dedicated.[77]

The Mauna Kea 88-inch facility had some unusual features due to its extremely high-altitude site. The coudé and control rooms were fed with oxygen-enriched air. The observing deck, including walls, was temperature-controlled by circulating water-glycol in order to prevent confusing the infrared instruments with heat radiating from inside the observatory.

The Hawaii and Texas telescopes included features designed to make them valuable for planetary astronomy, particularly spectroscopy (see figures 3.6–3.7). First, both were of fairly large aperture, located at sites with excellent seeing for photographic and spectroscopic studies; Mauna Kea offered a high and dry site for infrared observations as well. Both were situated at low northerly latitudes, so that the ecliptic, and particularly Mars at its favorable oppositions, would be high in the sky. Texas and Hawaii were at longitudes sufficiently separated from other observatories that bad weather or high water vapor would not likely ruin all observing opportunities for a favorable apparition. The spread in longitude also allowed meteorological features such as clouds or dust storms to be observed for longer periods. As a planet set at one observatory it could be picked up by another.[78]

Both were equipped with several Cassegrain foci (at the rear and sides of the instrument), and an optical train which could divert the light to a coudé room below. The Cassegrain foci accommodated fairly lightweight focal instrumentation, which could be left in place prealigned and precollimated, and could be rapidly selected by rotating a flat mirror inside the telescope. The coudé rooms for both contained a large, multipurpose coudé spectrograph with many possible combinations of gratings and cameras. Additional heavy focal instrumentation, such as interference spectrometers, could be located in the coudé room and selected by rotating a diverting mirror. The coudé room for both telescopes was basically a huge, walk-through optical bench, rigidly attached to the main piers for stability, and able to accommodate a wide variety of heavy spectroscopic and other equipment. Even the coatings on the secondary optics were chosen to allow flexibility. Aluminum has poor infrared response, while silver has fairly poor ultraviolet response. Depending on the spectral range of interest, the observer could select the appropriate diverting mirrors for maximum response in that range. Of course the primary mirror could only be aluminized, but significant gains in spectral response could be had just on the basis of the diverting mirrors alone.[79]

These two were the major optical telescopes built by the NASA planetary astronomy program, and almost from the instant of first light for each, they began to repay their investment in valuable and exciting results. But ironically enough, by the time they saw first light the Voyager Mars program, which had been their initial justification, was just a memory. It had been killed in the late summer of 1967, amid Apollo troubles, domestic unrest, the Vietnam War, and a host of other problems. For a few months, NASA was without any planetary program at all. But by then, while the Texas and Hawaii telescopes were being built, the ground-based planetary astronomy program had grown from a temporary solution to the needs of Voyager, into a diverse ongoing program with a more or less secure place within the planetary division at NASA. A host of other optical and

FIG. 3.6. The 107-inch reflecting telescope at the University of Texas McDonald Observatory was the largest optical planetary telescope funded by NASA. Photo courtesy of NASA.

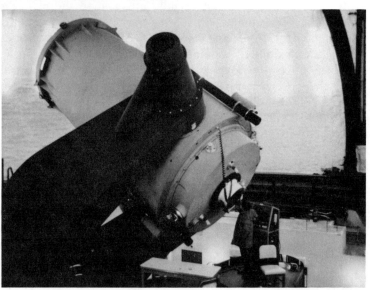

FIG. 3.7. The 88-inch reflecting telescope at the University of Hawaii's Mauna Kea Observatory inaugurated the use of Hawaii as a site for astronomy. Photograph courtesy of NASA.

radio/radar facilities had been built or improved and the research support had diversified. Between the demise of Voyager in 1967 and the maturation of the two major telescopes in 1969–70, a series of spacecraft had evolved, aimed at a more balanced exploration of the solar system. Mars still held its allure, but the planetary program, ground- and space-based, had taken on a broader agenda.

FOUR

The Program Matures (1965–1970)

While the needs of Voyager Mars assured the beginning of the NASA planetary astronomy program, a combination of factors assured the program's continuation, its integration into NASA's planetary exploration program, and its diversification to include radio and infrared facilities and studies of solar system bodies other than Mars. First, the planetary flight program broke out of its fixation on Mars and acquired (at least in some segments of NASA) the status of a post-Apollo priority, aimed at exploring the entire solar system. Second, the ground-based program earned the strong endorsement of advisory bodies within and outside of NASA, which also urged its expansion. Third, a host of factors constrained the ambitious planetary exploration plans and a much-reduced planetary exploration program emerged. This made ground-based surveillance all the more important, much in the way the missed opportunities of the early 1960s had assisted the ground-based program of those years. But a shortage of funds in all areas of NASA meant that more and more was being required of the ground-based program at the same time that funding for the program barely kept up with inflation.

OPTICAL TELESCOPES AND OTHER RESEARCH FACILITIES FOR PLANETARY ASTRONOMY

JPL AND TABLE MOUNTAIN

With the loss of the Caltech telescope, the JPL Optical Astronomy Group was left with an observatory equipped only with a small reflecting telescope. In the spring of 1965, as Hawaii emerged as the favored western site for a large telescope, JPL was allowed to add a 24-inch reflector to Table Mountain. But it was to be an unusual instrument. Ray Newburn and others at JPL wanted to do photoelectric photometry and especially high dispersion spectroscopy. James Gunn, a Caltech graduate student who later joined the JPL group for two years after his degree in 1966, suggested to Newburn that even if they couldn't get the large telescope they wanted, they could still put a coudé spectrograph on a small telescope. There would be less spatial resolution and scale with a smaller telescope, but high-dispersion planetary spectroscopy could still be done, since the spectrograph slit would be filled by the extended image of the planet.

In late fall 1964 JPL submitted its request for the instrument, and although Newell initially approved it, Deputy Administrator Dryden objected and wanted it reviewed in terms of NASA policy for ground-based astronomy support. In May 1965, after some convincing, Bill Brunk recommended approval of a minor construction request for Table Mountain for an operational Astro-Mechanics 24-inch reflector with both Cassegrain and coudé foci, a foundation, building, and dome all for under $100,000. Anything more than that would have required explicit congressional approval through a "construction of facilities" bill.[1] The digging out of the coudé room was approved by Brunk in April 1966, and approved by Newell at the same time but only on the condition that the coudé spectrograph itself be approved. The construction of the coudé room was approved in August 1966, and the spectrograph approved in July 1967.[2] Thus the acquisition of a second instrument on Table Mountain, an unusual combination of a 24-inch Cassegrain reflector with coudé room and coudé spectrograph, took almost five years from proposal to full installation. In part this was due to the fact that, in order to toe the NASA policy line on support to ground-based astronomy, acquisition had to proceed by stages. But the Table Mountain observatory also seemed to lack headquarters enthusiasm and full support even at JPL.

The JPL optical astronomy group continued to try to develop Table Mountain into a full-fledged observatory. JPL proposed a four-stage development plan extending over five years. The first stage proposed a fourier (Connes-type) interferometer, a Fabry-Perot interferometer, conventional coudé spectrograph, and an assortment of other instrumentation for the 24-inch. The second stage would be a cooperative program at McDonald using the 82-inch and the 107-inch with JPL trading use of some of its own instruments (fourier interferometer and radiometers) for a block of telescope time. The third stage would be a joint JPL–University of Washington 84-inch with coudé spectrograph. The interferometers used in the Texas program would then be transferred to the Table Mountain 84-inch. The fourth stage would be an infrared light bucket of 200 to 400 inches in diameter to feed a fourier interferometer.[3] This plan did not go over well at headquarters; due in part to administrative and policy problems, this proposal to develop Table Mountain was never implemented.[4]

It was as close as the JPL astronomers ever came to getting the instrumentation they wanted at Table Mountain. While a large planetary telescope at JPL would have made sense given the lab's concentration on the instrumented lunar and planetary exploration portion of the NASA program, several factors argued against JPL as the site for a ground-based planetary facility. First, there was the NASA policy of placing such instruments at universities where they could be of use to the astronomical community in general and could contribute to departments that had already demonstrated planetary interest. Second, throughout the period during which JPL was attempting to get a larger telescope and develop Table Mountain into an active observatory (1960–68), the lab was involved in three major lunar programs (Ranger, Surveyor, and Lunar Orbiter), the planetary flight program (Mariner, Voyager), and the worldwide deep space tracking network which maintained communications with all of the instrumented spacecraft and would be the link to the Apollo lunar landing craft. There was a feeling at headquarters and in the congressional committees that JPL was already overextended in these activities. In fact, relations between headquarters and JPL during this period were generally strained between Washington's desire to see JPL consolidate and contract out more of its work and the lab's tendency to do as much as possible in-house while taking on still more ambitious projects.[5]

The JPL astronomers and other staff had been at the center of the coalition described in chapters 2 and 3. Their arguments and admonitions regarding the role of planetary astronomy had been amply confirmed, and a system of integrated space- and ground-based study of the planets very similar to what they had been advocating was in place. But they didn't manage to complete the system exactly as they would have had it. Strained and competitive relations with NASA on the one hand and Caltech, the operator of the lab for NASA, on the other kept JPL from developing the major facility it hoped for and in a certain respect deserved.

The cooperative JPL–Texas activities did get going, however, and produced many valuable results. The Table Mountain 24-inch, once it was equipped with the proper spectrographic equipment, formed a valuable adjunct to the work being done on larger telescopes. The instrument was used in the development and testing of ground-based and spacecraft-borne planetary instruments, notably the Connes and other interferometers. The 24-inch found a valuable role to play in synoptic spectroscopic observations. Whereas the larger instruments could be used for spectrographic observations where high spatial resolution was required, the JPL instrument could be assigned to problems (such as the variability of water vapor content in the Martian atmosphere and the variations in the sulphur and sodium torus around Jupiter) where spatial resolution was less important than was a synoptic record to determine variations over a long time period.[6]

Mount Wilson and Palomar—Horse Trading

One way to gain more access to the large telescopes on Mount Wilson and Palomar was to help them in modernizing the instruments in return for some NASA access. In August 1965, Brunk met with Horace Babcock and Bruce Rule to discuss modernizing the Mount Wilson telescopes and making some modifications so that planetary observations could be done more efficiently and with less disruption to stellar and galactic observations.

They first discussed reactivating the coudé focus of the 60-inch. This included a flip-flop cage to hold new Cassegrain and coudé secondaries so that one could switch from Cassegrain to coudé focus quickly, a drive mechanism for the coudé flat mirror, and modernizing the right ascension and declination drives and readouts. Brunk inquired about similar modifications to the 100-inch, but Babcock wanted to see how the 60-inch worked out first.[7]

In addition to upgrading the Mount Wilson 60-inch, NASA also funded design studies and the primary mirror blank for a new 60-inch photometric reflector on Mount Palomar. Robert Leighton, Bruce Murray, Guido Munch, and others continued to use and improve the 200-inch and the 100-inch telescopes for planetary spectroscopy and radiometry under other funding from NASA.[8]

LOWELL OBSERVATORY—OLD DIRECTIONS, NEW VIGOR

Lowell Observatory had a long history of financial troubles since the death of its founder in 1916 (see chapter 1). With the arrival of William Sinton in 1957 and John Hall as director in 1958, however, things were looking up for the institution. In 1963 NASA had provided support to Lowell for planetary astronomy, enabling the observatory to assemble collections of planetary photographs from all over the world as one of two planetary data centers established earlier by the International Astronomical Union. Later NASA provided a facilities grant for a building to house the Planetary Data Center, which opened in May 1965.[9] In October 1965, the funding was expanded when William Baum resigned from Mount Wilson and Palomar to take over the Planetary Data Center activities.[10]

The copying of photographs at the Planetary Data Center and the various studies of the photographic records undertaken by the investigators there had alerted Hall and his staff to the serious gaps existing in the records. Hall had reported to the Space Science Board in 1966 that there was need for an international planetary patrol. Although there had been "sporadic programs of this sort" sponsored in the past, the different techniques, different filters, and different telescopes used had made the resulting patrol photographs difficult to analyze as a serial record. During the previous two such programs, at the favorable 1954 and 1956 oppositions (see chapter 2), "the staff of only two observatories took the pains to calibrate their plates." The coverage in the voluntary programs was so spotty, that examples of "yellow cloud" motion could be identified on only three of the entire collection of plates.[11]

By early 1968 the international planetary patrol had taken on a new significance. Advocated by the JPL and other groups from 1959 to 1962, the merits of the program had not been recognized by headquarters officials. But after analysis of the collections of photographs at the Lowell center, and after it became evident that Mars lander needed such information as it could provide, the planetary patrol gained welcome attention and support.

The proposed International Planetary Patrol would lease existing 24-inch–class telescopes from several observatories around the world, and build two such instruments at Cerro Tololo and Mauna Kea. Cameras and barlow lenses were provided for each telescope to produce images of the same scale, equivalent to Clyde Tombaugh's f/75 design. Special uniform color filters and other equipment would be loaned to the participating observatories, and a local observer hired especially for the patrol. Series of photographs of Mars, Venus, and Jupiter would be taken at preselected times, using identical filter combinations in sequence, with data concerning the color filter used, date and time of photograph, and photometric calibrations automatically recorded on each frame. The exposed 35-mm motion picture film would then be returned to Lowell for developing and copying. In that way Lowell could control the developing process and ensure uniformity and comparability of the photographs from various stations. The formal arrangements removed the problems of cooperation which had plagued earlier planetary patrol efforts. The proposal carried the endorsement of NASA's Lunar and Planetary Missions Board.[12]

The patrol got underway for the 1969 Martian opposition (see table

TABLE 4.1
International Planetary Patrol Telescopes (as Proposed).

Source: Hearth to Naugle, op. cit.

*Lowell Observatory Flagstaff, Ariz.	24-inch Lowell refractor f/16
	24-inch Morgan reflector f/104
*Mauna Kea Observatory Mauna Kea, Hawaii	24-inch reflector f/18
*Mount Stromlo Observatory Canberra, Australia	26-inch Yale-Columbia refractor f/16
Kodaikanal Observatory Kodaikanal, India	24-inch reflector f/8
Republic Observatory Johannesburg, South Africa	26.5-inch refractor f/16
*Cerro Tololo, Chile	24-inch reflector f/75
*New Mexico State University Park, N. Mex.	24-inch reflector f/75

*participant in 1969 opposition planetary patrol

4.1). In two years, the patrol efforts had tripled the collection of photographs which Lowell had previously assembled from existing files, and the planetary data centers had more than 40,000 sets of photographs cataloged by computer. Since examination of earlier Mars maps had shown errors as large as five to ten degrees in the locations of features, the Lowell group made new albedo maps of the planet from scratch. Using a projection device that merged full disc photographs of the planet with an appropriate latitude and longitude grid, airbrushed mercator projection maps were prepared for NASA mission planning and for other studies at Lowell.[13]

The patrol photographs were used by a variety of investigators, from both the planetary research center and other institutions. Baum, together with L. J. Martin, made an extensive study of existing prepatrol photographs of Mars which showed clouds that were identifiable on successive nights. They used the projection device to translate the photographed cloud positions onto maps, and so compiled "cloud histories" from which they were able to estimate a mean velocity of cloud motions of 5.6 km/hr, much lower than estimates other observers had made based on visual observations. They also noticed a tendency of the clouds to "avoid" the dark areas of the planet's surface.[14]

In addition to the above activities, the NASA planetary astronomy program continued to support a wide variety of work at various institutions in all spectral regions (see appendix 5).

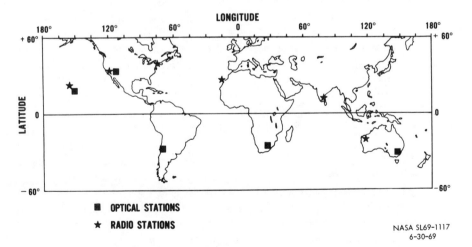

PLANETARY MONITORING PATROLS
OBSERVING NETWORK

FIG. 4.1. The International Planetary Patrol network of observatories could follow rapid changes occurring on the planets. Before the planet set at one station it could be picked up by the next one west. Graphic courtesy of NASA.

THE HALL REPORT ON PLANETARY
ASTRONOMY—RISING VOICES

The Space Science Board had recommended the formation of a panel on planetary astronomy at its 1962 Summer Study, suggesting Lowell Observatory director John Hall as chairman. It repeated this recommendation at the 1965 Summer Study, where the post-Apollo goal of solar system exploration was fleshed out and the needs of ground-based planetary astronomy were discussed at length. The failure of the Whitford committee to consider the needs of this specialty moved the board to action. With NASA support, the ad hoc Panel on Planetary Astronomy was established under Hall in 1966. The panel was charged

> to evaluate the current state of knowledge in planetary astronomy; to indicate fields of ground-based astronomy likely to be particularly productive in the future; to assess and compare investigative techniques now in use or under development, and to make specific recommendations on present and projected requirements for personnel, personnnel training, and new or improved facilities.[15]

The Hall Report was to be planetary astronomy's version of the Whiteford Report: a document which sought to convey to policy makers and funders the present state, importance, and needs of a field, to define its boundaries and show how it related to other similar but separate activities. Essentially the same forces motivated by the reports: in an age of both government largess and government demands on science, it was important for scientific constituencies to proclaim just who they were, what knowledge and services they could and could not provide, what they needed in the way of equipment and personnel to provide those services, and how and to what extent those needs were likely to be met. It is significant that planetary astronomers and planetary scientists felt they had to proclaim their identity and needs with their own report, separate from the Whitford Report, which billed itself as covering the next ten years of "ground-based astronomy." It is also significant that they did so under the auspices of the National Academy of Science's Space Science Board, rather than under the auspices of the academy's Committee on Science and Public Policy (the body that had sponsored the Whitford Report on astronomy). Clearly, the identity of ground-based planetary astronomy was more closely allied with spaceflight than with mainstream astronomy.

The Hall Report first tried to place ground-based planetary astronomy in perspective—to show how it could contribute to the overall goal of understanding the origin and evolution of the solar system, and to weigh the relative merits and disadvantages of ground-based and space-based techniques. Certain ground-based observations played a supporting role in the planning and execution of spacecraft missions to the planets, the results of which were sure to surpass in quality and quantity the ground-based methods. But there were areas in which ground-based planetary astronomy was the only foreseeable away to obtain certain data, and areas in which ground-based observations would be superior to probe-based observations. Much as the Whitford Report had argued that space-based techniques would in no way totally surpass and make obsolete ground-based astronomy, the Hall report argued for the importance of planetary astronomy on it own merits as well as in its probe-supporting role.[16]

Four chapters of the report then examined in some detail the present state of knowledge, active questions, and relevant techniques for the study of planetary dynamics, surfaces, atmospheres, and interiors. The panel found the standard techniques of photometry, polarimetry, spectroscopy, and radiometry still valuable in determining structure, composition, and other characteristics of planetary surfaces and atmospheres.[17] Some newly improved techniques, particularly high-resolution infrared and radar mapping, were quite promising, if only observation time on large instruments were available. The calibration of earth-based techniques by *in situ* and laboratory measurements promised to make the former much more reliable. But single or restricted series of observations could not be the last word on planetary surfaces, since many instances of periodic and secular changes were known. This called for long-term monitoring or patrol type programs, which were only feasible as ground-based efforts and required instruments dedicated to planetary astronomy.[18]

The authors reviewed the current state of knowledge of planetary atmospheres, suggesting many particular areas in which improvement was called for. To detect more than the simple presence or abundance of an element, to analyze the distribution of constituents and changes over time, required considerable study. An approach combining a great deal of basic observational data from all of the ground-based methods promised advances in unraveling the structures of planetary and cometary atmospheres. The problem was not one of theoretical synthesis, but rather the need for very basic observational data, without which theories could not even begin to be constructed.[19]

Chapter 6 discussed observational techniques and facilities. Several new techniques had been developed which were not generally known beyond planetary astronomy specialists, and the authors thought it advisable to introduce them. Among the optical techniques, Hall had been most impressed in 1965 by the Connes interferometer. The device was possible only because of modern computing techniques, which were required to convert the output of the interferometer into synthetic spectra. So high was the resolution and efficiency of the new interference spectrometers that their spectra showed many lines that had never been produced in the laboratory, thus calling for additional laboratory spectroscopy to try to identify them.[20]

The seventh chapter examined the state of graduate training in the planetary sciences and "whether the influx of new blood is adequate to support an imaginative space program." The authors presented a consider-

able number of statistics from a survey conducted for the report, estimating
that not more than a hundred Ph.D. theses in planetary science had been
written in the period between 1960 and 1967. The authors examined the
question of where within the universities and scientific societies such study
should properly be undertaken.[21]

The report closed with a list of primary recommendations, and a corre-
sponding list of additional needs for planetary astronomy and planetary
science (table 4.2).

The Hall Report was far less at odds with NASA and its activities
than many previous and later SSB reports. Its survey of the use of optical
telescopes revealed very low percentages for planetary astronomy, but
having seen the existing NASA planetary astronomy program of instrument
development, particularly the instruments under construction at Texas and

TABLE 4.2
Recommendations of the Panel on Planetary Astronomy.

Recommendations

1. A 60-inch telescope designed for spectroscopy, interferometry photography, and
   photometry of the planets at a superior Southern Hemisphere site (preferably
   Chile).
2. Maximum use of existing radar facilities, and construction of large filled-aperture
   facilitiy with capability for planetary radar astronomy.
3. 120-inch infrared light bucket with coudé focus at an exceptionally dry site
   for photometry and spectroscopy.
4. 36-inch infrared telescope to be installed in the NASA Convair jet and made
   available to qualified users for high-altitude infrared studies.
5. Several existing and projected radio telescopes to be made available for planetary
   studies.
6. Development of fourier interferometers, with one rugged enough to be installed
   in an aircraft.
7. Development of radiation detectors in all spectral regions, image tubes, and
   devices and techniques for improving the quality of images.
8. International Planetary Patrol, distributed in longitude, 1969–1974.

Additional Needs Cited

A. Southern Hemisphere astrometry of comets and asteroids, automation of measuring and data-handling.
B. More effort to recover, analyze, and catalog meteorites and extraterrestrial matter.
C. Additional rocket flights and use of the Orbiting Astronomical Observatory for
   ultraviolet planetary observations.
D. Systematic and coordinated facilities and program of laboratory spectroscopy
   in all wavelengths, calibrated with returned samples of extraterrestrial matter.
E. Strengthen graduate training in interdisciplinary planetary sciences; establish
   national society or division of existing scientific society devoted to planetary
   science.

Hawaii, the panel pronounced, "Except for the occasional need for large blocks of time. . . . Ground-based observational requirements in the Northern Hemisphere appear to be met by existing telescopes. Additional needs in the near future could be filled at least in part by the two major instruments now under construction, and by an inexpensive large telescope recommended [the infrared light bucket]." However, the panel repeatedly called for the construction of a 60-inch planetary instrument in the Southern Hemisphere, preferably in Chile. The NASA program substantially supplied the vast majority of its desiderata, and made serious attempts to supply the rest. In October 1967, while the Hall panel was writing up its findings, Brunk had made a list of proposed "seed grant" projects for his program. First on the list was obtaining higher power transmitters for existing radar antennas (Arecibo and Haystack). Second was establishing the worldwide planetary patrol. Third was obtaining more Connes interferometers. Fourth, constructing additional centimeter- and millimeter-wavelength radio antennas.[22]

In September 1969, with the Texas telescope completed, the Hawaii instrument nearing completion, and the planetary patrol underway, Brunk listed the 60-inch Cerro Tololo telescope second behind the resurfacing of the Arecibo radio telescope's antenna and provision of a radar transmitter as fiscal year 1971 priorities. The large infrared telescope recommended by the Hall panel was third. Brunk noted in the backup material for this request that "an effort was undertaken to support the construction of a 60-inch planetary telescope [at Cerro Tololo] in time for the 1971 opposition of Mars [as recommended by the Hall panel] but this was not possible within existing budgets." Brunk hoped to get $1 million to start on the instrument and another million to finish it in time for the 1973 opposition, but that too fell through.[23]

Site surveys began in 1969 for the large infrared telescope, which, after a particularly troubled and protracted development, appeared on Mauna Kea in 1979 as an exceptionally fine 3-meter instrument. Frank Low and Gerard Kuiper had already been experimenting with infrared observations from a Learjet and a Convair CV 990. The full airborne 36-inch infrared facility, after feasibility studies starting in 1969, was established at Ames Research Center in a Lockheed C 141 and named the Kuiper Airborne Observatory in 1975, after the astronomer's death. The planetary patrol started operation in April 1969, managed by Lowell Observatory, and contributed in significant ways to all later planetary missions. As the Mars requirements lessened in the early 1970s, more work on outer planets, comets, and asteroids was supported. Some use was made, under the guest investigator program, of the Orbiting Astronomical Observatory and the International Ultraviolet Explorer satellites for ultraviolet planetary astronomy, but the design of the spacecraft made them difficult to use for planetary work. Laboratory spectroscopy was already being supported at Ames Research Center, JPL, Goddard Space Flight Center, and several universities. Hence, most of the recommendations of the Hall panel were

FIG. 4.2. The 3-meter Infrared Telescope Facility (IRTF) at the University of Hawaii's Mauna Kea Observatory fulfilled one of the Hall Report's major recommendations. It was the last major planetary telescope established by NASA. Photograph courtesy of NASA.

in one way or another implemented by NASA over the ensuing decade. Planetary astronomers acquired access to a powerful and varied set of instrumentation.

THE PLANETARY FLIGHT PROGRAM

Planetary astronomy at NASA had initially gained a strong foothold because the mammoth Voyager Mars program had been jeopardized by inadequate planning data. A protracted and complete overhaul of the planetary exploration program, including its scientific rationale, provided opportunities for a host of internal and external advisory and planning bodies to become aware of the potential of ground-based study.

With no Mars missions scheduled until 1969 and 1971, and Voyager to fly no earlier than 1973, any information on the Martian atmosphere and surface likely to be integrated into the lander design would have to come from ground-based study. Some modifications could be accommodated in light of data from Mariner Mars 1969 and 1971, but any gross revisions would likely upset the 1973 mission.

In June 1967, as a spare Mariner V was being readied for a somewhat impromptu launch to Venus, Congress made deep cuts in NASA's budget. A disastrous fire in January had killed three astronauts on the launch pad during a test of the Apollo spacecraft and launch vehicle, and the aftermath included shakier relations between NASA and Congress, as well as the transfer of money from other NASA programs into Apollo to recover from the setback. The Vietnam War was escalating, and budgets were generally getting tighter. Public support for the space program was eroding.[24]

By August, Voyager was in deep trouble in Congress. Early that month the Manned Spacecraft Center in Houston sent out to industry a fairly routine request for proposals for studies of manned missions to Mars and Venus. But in this case the timing was most inopportune. Voyager seemed then like a "foot in the door" for extravagant manned Mars missions, which would end up costing much more than the already troubled Apollo program. Congress rebelled, eliminating all funds for Voyager and for Mariner Mars 1971.[25]

In *Science,* Luther J. Carter noted that 1967 was to have been the year of post-Apollo commitment, but "this year has been one of debacle for NASA's plans to mount, for the 1970s, a major program of unmanned planetary exploration." With only one reworked spare Mariner heading for Venus, two Mariner Mars flybys scheduled for 1969, pressure from the Space Science Board and from the President's Science Advisory Committee for a viable planetary exploration program, and "Voyager" a word that could hardly be uttered before congressional committees again, NASA planners went back to the drawing boards.[26]

As guidelines for planning on a virtual blank slate (but by no means with a blank check), NASA had the SSB recommendations that planetary exploration be the major post-Apollo goal, using a combination of spacecraft and ground-based observations. The PSAC had also given NASA its advice on this matter, in a report prepared during 1966 by its space panels. PSAC agreed with the SSB that planetary exploration, leading ultimately to manned voyages, should be the major effort of NASA in the post-Apollo period. It also agreed with the SSB that "opportunities are great for obtaining detailed information about planetary surfaces and atmospheres from earth-based optical and radar observations and from earth-orbital optical observations," and explicitly recommended "the determination of the appropriate ground-based and earth-orbital observations required to insure the maximum effectiveness of direct exploratory ventures projected for the 1970s."[27]

PSAC had also been somewhat disturbed by the apparent "absence of integrated NASA planning in this area," since it had received from the agency two different plans for the exploration of Mars. One, from Newell's Space Science Office, presented the Voyager program. Another, from his rival George Mueller's Office of Manned Space Flight, presented plans to use modified Apollo hardware for a piloted Mars flyby, perhaps by 1975, which might "return a greater amount of useful information than is likely to be returned by the entire proposed program of Voyager missions to Mars in the 1970s."[28]

In both proposals, PSAC found fault with "the implication that sufficient knowledge exists at present to warrant the commitment of large resources for the exploration of a single planet, Mars, at the expense of more balanced efforts for the other planets." Noting that the discovery of water vapor on Venus (by ground-based work) generated a "renewed interest in that planet as a possible locale of organic molecules, if not primitive life-forms," the committee recommended reconnaissance of Venus and Mars before deciding which merited more intensive investigation. PSAC also put exploratory missions to Mercury and Jupiter at "almost as high a priority."[29]

The SSB had, in the previous two years, somewhat moderated its emphasis on Mars. In July 1966 it had told NASA Administrator Webb that "the Board finds nothing critically new that might modify the 1965 study although the Board views Venus with somewhat greater interest than earlier, ranking it thus close to Mars and believes that significant scientific objectives can be achieved by a fresh and economical program of a number of small planetary probes directed to the nearer planets, including Jupiter."[30] In 1967, SSB member and proponent of economical small spacecraft James Van Allen had chaired an ad hoc committee on small planetary probes, which had strongly recommended the use of smaller systems for planetary exploration to pave the way for Voyager-class spacecraft.[31]

Thus when the NASA managers met in the fall of 1967 to take a fresh look at the planetary program, they had endorsement from both the SSB and PSAC for a program in which Mars would enjoy high priority, but which had to give some attention to the other planets as well. The SSB and PSAC had also strongly endorsed the ground-based planetary astronomy program, and NASA was aware that in the fiscal austerity of the times, every planetary mission had to be designed to do something that could not be done from the ground, and would have to rely on ground-based results rather than an extra probe or two to help define the planetary spacecraft and their experiments. The spaceflight missions they would plan, moreover, would have to be more economical and more restricted in scope than they might otherwise have envisioned. All of these forces operated to ensure the continuing importance of the ground-based planetary astronomy program, but also worked against it in limiting its potential funding. With a congressional ban on new program starts, advanced planning studies would have to be funded out of Supporting Research and Technology funds, the same pot that financed the planetary astronomy activities.[32]

During September and October, as Voyager died in conference committee, NASA officials met with Webb to map out a planetary exploration program which would combine the various (sometimes conflicting) desiderata mentioned above. The support of the scientific community was important, since it would be an uphill battle in Congress.

On November 8, Webb replied to a question from Senator Clinton Anderson on what plans NASA had for planetary exploration. Luther Carter thought the question "inevitable," and that "Webb must have felt that his proposals might gain momentum if they became known to space scientists, contractors, and other interested parties who could apply political pressure." Webb proposed a "bold" plan for five Mariners between 1971 and 1976: two Atlas-Centaur–boosted modifications of existing Mariner equipment to be used as Mars orbiters in 1971; Titan IIIC–boosted missions to Venus in 1972 and 1973, the latter taking a gravity assist from Venus and continuing to Mercury; two Titan IIIC–boosted Mars orbiters with small entry probes for 1973; two Saturn V–launched heavy Voyager-type combination orbiter-landers for 1975. The alternative, Webb told the senators, would be to let the planetary exploration teams and capability built up over the years dissipate.[33]

While the budget was being thrashed out, NASA requested another summer study from the Space Science Board to reevaluate the planetary exploration program in light of the new scientific knowledge gained since the 1965 Summer Study, and particularly in light of the new economic realities. Twenty-two consultants chaired by Gordon J. F. MacDonald met during the week of June 10, and presented their findings to NASA on June 16. The full SSB endorsed the study recommendations immediately thereafter.[34] The SSB recommendations were helpful in endorsing planetary exploration efforts, but "somewhat detrimental since they did not coincide exactly with the agency's announced goals. In times of extreme congressional scrutiny, Webb and his colleagues at NASA would prefer a more closely orchestrated variety of advice."[35]

The board found the expenditure NASA had slated for planetary exploration (less than two percent of its budget) "totally inadequate," and recommended that a "substantially increased fraction of the total NASA budget be devoted to planetary exploration." They recommended that the resources "currently intended for support of manned planetary programs be reallocated to programs for instrumented investigation of the planets."[36]

NASA's ground-based planetary astronomy program drew consistent praise from the board. They commended NASA on the construction of the three large telescopes (Kuiper's 60-inch, Texas, and Hawaii), but recommended a Southern Hemisphere intermediate-size optical telescope, a large infrared telescope, and further development of infrared detectors and high resolution interferometers. They also commended NASA on the International Planetary Patrol, but were concerned that the enormous amount of data accumulated would not be promptly analyzed, and so recommended that NASA further support the analysis of the data by additional investigators.[37]

In addition to the optical range, the board was most impressed with new developments in radio and especially radar astronomy. In the years since the 1962 Summer Study, radar investigations had revealed the sense and rate of rotation of Venus and Mercury, and shown them to be coupled in resonance. The first accurate determination of the radius of Venus by radar, combined with Mariner 5 results, had shown that the Soviet Venus 4 probe did not transmit its measured temperatures and pressures from the surface, but from a considerable altitude, thus revising the values from 550/ K and 20 atmospheres as reported by the Soviets to 700/ K and 100 atmospheres. With improved sensitivity and high-power transmitters, radar astronomy would be able to map the surfaces of Mercury, Venus, and Mars, and by reflecting signals off the Galilean satellites of Jupiter and the satellites of Mars near occultation, could probe those planets' atmospheres. Accordingly, the board recommended that NASA "fund the development and operation of a major new radar observatory," with a sensitivity improved a thousandfold over existing systems, and estimated the cost around $30 million.[38]

The SSB also saw great potential in earth-orbiting astronomy satellites (the Orbiting Astronomical Observatories, Small Astronomy Satellites) and urged that use of these satellites be pursued vigorously, especially for ultraviolet spectroscopy and high-resolution imaging. The board also urged the use of the infrared aircraft telescopes under construction.[39]

In all, the portion of NASA's planetary exploration program that drew the highest marks from the study was planetary astronomy, but the SSB still saw a need for expansion and diversification. The same sort of remarks figured prominently in the report of the SSB Panel on Planetary Astronomy, released at about the same time (see above).

In addition to endorsements from the President's Science Advisory Committee, the SSB, and other such external advisory bodies, the NASA ground-based planetary astronomy program also enjoyed the support of other, internal NASA advisory bodies. Two special missions boards had been established in 1967, in part due to recommendations from an ad hoc advisory committee established to examine relationships between NASA and the scientific community. The Astronomy Missions Board and the Lunar and Planetary Missions Board were established to bridge the gap between outside advisory scientists, such as the SSB, and inside committees such as the Space Sciences Steering Committee.[40]

THE LUNAR AND PLANETARY
MISSIONS BOARD

The Lunar and Planetary Missions Board (LPMB), chaired by John Findlay, included a prestigious cross section of the various disciplines represented in planetary science, and included astronomers Thomas Gold, Irwin Shapiro, George B. Field, P. Goldreich, F. S. Johnson, M. J. S. Belton, and John Hall. Although it never published a comprehensive plan or set of working papers as the Astronomy Missions Board did, the LPMB approached the lunar and planetary program as an integrated whole. It wished to avoid "a series of unconnected projects designed only to capture the attention of Congress or the public." It was in favor of balance, order, and careful planning, as well as flexibility to change mission designs and experiments when new discoveries called for it.[41] A prominent place in planetary exploration for ground-based astronomy appealed to the LPMB's sense of order and balance. The need for NASA to enter the field was a foregone conclusion for many members of the LPMB, themselves planetary astronomers and accustomed to the frustrations of attempting to get telescope time for planetary work. They worked closely with Voyager program personnel, and with Brunk in planning the extensive Mars observations on which Voyager lander designs depended. But the congenial relations between NASA and the LPMB were strained to the breaking point when the Voyager program was cancelled in the fall of 1967. While the LPMB was grudgingly willing to accept the cut of Voyager by Congress, the substitute program which Webb proposed in November went against its recommendations for more balance and diversity in planetary targets, and use of smaller probes in place of large, expensive, complex systems like Voyager.[42]

THE ASTRONOMY MISSIONS BOARD—1967–1970

The Astronomy Missions Board (AMB) was the astronomy counterpart to the Lunar and Planetary Missions Board, established in the aftermath of the Ramsey Report to strengthen NASA ties with the academic scientific community and aid in agency planning. The AMB, which reported to Newell and was chaired by Leo Goldberg, was composed of eighteen regular members, and panels with thirty-one additional scientists represented. In June 1968, the AMB asked each of its several panels to prepare a long-range space astronomy program for its particular subdiscipline, recommending the most exciting problems, an estimate of the space-based observations, and complementary ground-based observations required for progress, and the types and numbers of instruments needed for such a program.[43] The planetary panel confined its recommendations to "planetary space astronomy": observations of the planets which could be made from sounding rockets, aircraft, small astronomy satellites, and a large space telescope.[44]

The AMB's Ground-Based Astronomy Working Group noted that space astronomy and ground-based astronomy feed on one another, and that the "existing ground-based observing centers and university astronomy departments are the principal source of manpower and the training ground for young astronomers needed in astronomical projects involving observations from space." In addition to providing young astronomers, ground-based astronomy was essential to the space program in helping select the "crucial observations that will yield the most useful return from the specialized instruments sent above the atmosphere," and in supplementing space-based observations through the versatility of ground-based equipment.[45]

Hence, the group concluded, "the stakes are very high, and NASA can hardly avoid being concerned about the health, vigor, and technological advancement of the astronomical community from which it must draw its strength. Nor can NASA avoid concern about the adequacy of the ground-based telescopes needed for optimum exploitation of space observations."[46] The group saw no problem with NASA supporting those areas of astronomy where it had the clearest mission needs, but warned that NASA must "remain cognizant of the need, and take steps to avoid letting the priorities be determined by jurisdictional rigidities [between NASA, NSF, and other funding agencies]." Regarding large optical telescopes, the group noted that the national observatory telescopes were presently oversubscribed by a factor of two, and that the number of astronomers was increasing at around fifteen percent per year. Even the McDonald and Mauna Kea Telescopes, which had guaranteed NASA a portion of their observing time, fell short of the needs of planetary astronomy.[47] The group also found the radio and radar facilities inadequate, citing the large number of discoveries and revisions of previous ideas concerning Mercury, Venus, and Jupiter made possible by radio/radar astronomy. It recommended that NASA support the construction and operation of new radio/radar facilities for planetary work, and work closely with NSF. The working group concluded with recommendations that NASA underwrite a 200-inch optical telescope in the Southern and Northern Hemispheres, a 400-foot steerable paraboloid for radio/radar astronomy, and a variety of smaller instruments.[48]

AMBITIOUS PLANS FOR PLANETARY ASTRONOMY AND EXPLORATION—1969

In all of the post-Apollo planning discussions various segments of the agency faced the problem of what to do with the enormous organization, facilities, and capabilities that had been built up for the lunar landings. The Saturn V production lines in industry, the Marshall Spaceflight Center which managed the vehicle, and the Cape Kennedy launch facilities, were all facing the prospect of simply running out of work to do after the Apollo missions.

The manned spaceflight side of the agency (Mueller's office) which managed these centers and facilities had long been planning lunar bases, large space stations, and manned missions to the planets, particularly Mars, as the next logical steps after Apollo.[49] The space science side of the agency (Newell and Naugle) had been moving in two directions. A large optical observatory, either on the moon or in earth orbit, had been debated since the 1961 SSB Summer Study, and the chief points of uncertainty seemed to center on the choice of the moon or earth orbit as a location, technological feasibility of the large mirrors, attitude and pointing control, data retrieval, and the role of man in the facility.[50] The other direction, which seemed to be favored by the SSB and other advisory committeees, was a vigorous program of planetary exploration. Here the debatable points were whether to use large probe systems like Voyager or small Pioneer-class spacecraft, the choice and sequence of planets to be investigated, and the proper role of people in planetary exploration.

President-elect Nixon came into office with two major concerns: Vietnam and the nation's faltering economy. To assess the options available for the space program, Nixon asked Charles Townes to assemble a panel of consultants to write a "transition report" for the new administration. The transition report, presented to Nixon in January 1969 but not publicly released, advocated a strong program of planetary exploration, opposed a large space station in favor of a space shuttle, and urged the dissolution of the manned/unmanned dichotomy which had been such a sore point at NASA over the years.[51] Nixon appointed an interagency Space Task Group, chaired by Vice President Agnew, to produce a report with recommendations for the post-Apollo space program by September 1969.[52]

The discussions and planning of the Space Task Group were characterized by the tensions between the Administration's desire to keep budgets down and NASA's desire to secure an imaginative and vigorous post-Apollo mission for the agency.[53] The growing enthusiasm in certain NASA circles for the commitment to a manned Mars mission, just after the successful Apollo landing, provoked intense criticism from Congress, the press, and the public. As the Space Task group wrote its report for the President's consideration, in the atmosphere of growing criticism of expensive new initiatives, it attempted to present a range of alternative plans of varying boldness and cost. But they were essentially a range of approaches to the same goal, manned planetary exploration, with the "first target" being man on Mars before the end of the century and perhaps as early as 1981. The President received the report on September 15, but did not respond for another six months.[54]

Although the President did not immediately respond, the scientists did. When the Space Task Group Report and the NASA submissions were published, they catalyzed a growing displeasure in the scientific community. Thinking that with the achievement of Apollo, science would at least assume its rightful place within NASA, many within the scientific community had looked for a balanced program in which science would predominate and the engineering equipment and techniques built up at such cost would be utilized to provide routine access to various areas of space for scientific use. What they saw in the NASA submission to the Space Task Group was another Apollo, only worse: a proposal for protracted, complex, manned extravaganzas which, although they tipped their hat to science, would squeeze scientific activities into a restricted "piggyback" role, and dominate the agency for another decade or more.[55] The reaction was so bad that on November 6, 1969, Newell asked his assistant to "assemble a file of all the letters from scientists complaining about different aspects of the NASA science program."[56] Two years later Newell described the situation in the following terms:

Relations with the Space Science Board, and also with our own Boards and Committees, began to come apart about the time the Space Task Group Report was published. Strains developed because the Boards and committees felt they were not being effective or listened to by NASA. The Budgets in the Space Task Group Report were regarded as appallingly high. The emphasis given to very large scale programs—space shuttle, space stations and space bases, lunar bases, nuclear shuttles, Grand Tours, and manned missions to Mars—had a very negative effect.[57]

Editorials appeared in scientific journals, letters arrived at NASA, and various advisory boards and committees began to express their displeasure. Newell prepared a briefing paper for NASA Administrator Paine, who had replaced Webb in the fall of 1968, in which he reviewed the concerns of the scientific community, those of the various advisory bodies, gave a sampling of complaints from individual scientists, and recommended a number of actions NASA could take to improve relations. "The dissension is serious," he wrote, "and requires the careful attention of NASA management. Science is and must continue to be an important objective of the NASA program."[58]

One of the complaints that repeatedly was expressed, Newell wrote, was that "not enough supporting research and technology funding goes to support spadework in the science area." In response to this, Newell recommended that NASA develop a plan for increased support to ground-based science, "including long-range supporting research and technology required to establish a broad basis for future space research." He particularly recommended increasing support to astronomy.[59] Other options for strengthening NASA's relationship with the scientists included finding a qualified, distinguished, and respected person to become "chief scientist" of the agency, reorganizing the advisory structure, and attempting to communicate to the scientific community the need for a long-range plan such as the Space Task Group Report and the notion that science would ultimately benefit from it. But the most important action, Newell thought, was to "take specific steps to meet valid concerns that underlie the expressed discontent," and "ensure that an adequate, high-quality science program is included in and supported by the NASA efforts."[60]

It was in this context that Newell approached the NSF to discuss an expanded program of NASA support to ground-based astronomy. In early November, Newell and Naugle met with NSF Director William D. McElroy, Astronomy Program Director Robert Fleischer, and others at NSF. Newell reviewed the evolution of NASA's policy of support to ground-based astronomy from its response to the early needs of the planetary flight program to the formalization of the policy after the Whitford Report, and stressed NSF's concurrence at the time in that policy. Looking to the future, Newell said that astronomy must be "a major element of the NASA science program," and foresaw that ground- and space-based activities would "meld together into a sound total program." Thus Newell thought it "appropriate and desirable for NASA to assume a greater share of responsibilty toward the support of ground-based astronomy," not just in planetary astronomy but in all areas. Newell was seeking NSF's views as to whether such an expansion would be appropriate. McElroy had no objections, provided the two agencies closely coordinated their plans, since "otherwise moves by one agency might generate potential responsibilities for the other." They agreed to establish a formal coordinating mechanism, with Fleischer from NSF and Henry J. Smith from NASA as primary contacts, and resolved that Paine would write to McElroy outlining the expanded NASA policy.[61]

On December 2 McElroy and the NSF astronomy program officers met with Newell, Smith, Roman, Goetz Oertel, and Brunk of NASA to discuss detailed coordination plans. The differences in philosophy between NASA and NSF became evident during the meeting. Fleischer "expressed deep concern" over previous experiences where agencies had established expensive facilities for mission needs and then failed to provide adequate operation support. NSF then had to pick up the operational expense for facilities "which may not represent, in the opinion of NSF and the scientific community, the best use of these funds for astronomical research." Fleischer thought that the needs of the astronomical community as a whole should be considered in the construction of any astronomical facilities, while some of the NASA people "felt that urgent mission requirements provided sufficient justification for the construction of new facilities." The groups exchanged their funding plans for fiscal year 1971. NASA had scheduled about a $1 million increase in its budget, with the upgrading of Arecibo for planetary radar, a 60-inch planetary telescope for Cerro Tololo, or a large infrared telescope as possible candidates for the increase, depending on NSF's final submission to the federal Bureau of the Budget and Congress.[62]

Paine's letter to McElroy followed on December 12, 1969. He reviewed previous NASA policy, and noted that the past level of NASA support "has not provided the facilities and operational support to meet our program needs." For that reason, and while recognizing NSF responsibility in astronomy, NASA planned to augment its support. Paine cited four reasons for the augmented support. First, the earth-orbiting solar and stellar astronomy satellites required more ground-based supporting observations; second, the "greatly increased planetary program . . . require[s] more continuous and intensive correlative ground observations than present observatories can provide"; third, astronomy in general was rapidly expanding and ripe for using earth-orbital astronomical observatories which would require additional ground-based research; fourth, a number of NASA advisory bodies, including the Space Science Board, Lunar and Planetary Missions Board, and the Astronomy Missions Board, had all urged a higher level of support to ground-based astronomy by NASA in order to more fully exploit the expanded planetary program, as well as the activities in space astronomy.[63]

Paine's overtures to the NSF promising greater and more diversified NASA support to ground-based astronomy were part of his attempt to get as large a NASA program for the post-Apollo period as he could, even in the face of many signals that NASA would have to adjust to a much more modest program.[64] In the meantime, as the fiscal year 1971

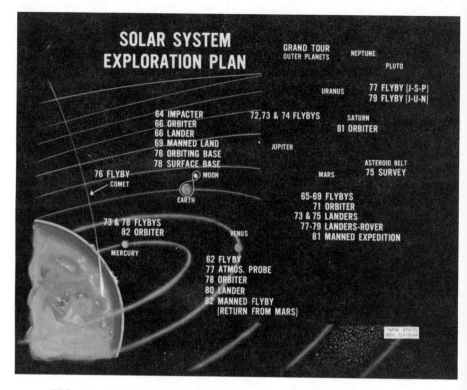

FIG. 4.3. This "Solar System Exploration Plan" of 1969 envisioned an ambitious series of planetary spacecraft. Photograph courtesy of NASA.

budget was being worked out, the Bureau of the Budget made severe cuts in NASA's request in the fall of 1969. The Christmas season, however, was to bring worse news for NASA's budget as a whole, and the space science budget in particular. As the struggle continued through the winter, NASA's budget was cut twice again, winding up at about $3.3 billion, twenty-five percent less than the original request and fifteen percent less than the previous year's budget. The remaining Apollo missions were stretched out, Saturn V production was suspended, and launch of the Viking Mars lander missions, the successor to the ill-fated Voyager program, was postponed for two years from 1973 to 1975.[65]

In March 1970 Nixon released a three-page statement. It endorsed none of the program options in the Space Task Group Report, spoke of the space effort in the most general terms, and emphasized the need to be both bold and also to attend to problems on earth. The statement spoke of the need to integrate space activities as a "normal part of our lives" and to avoid "a series of separate leaps, each requiring a massive concentration of energy and will and accomplished on a crash timetable." Nixon said, "We should move ahead with bold exploration of the planets and the universe," and that we would "launch unmanned spacecraft to all the planets of our solar system" during the next decade, eventually sending men to explore Mars as part of this exploratory effort but as a "longer range goal."[66]

Yet Paine, as Newell put it, "chose to focus on the president's call to be bold, rather than on his admonition to proceed at a measured pace. . . . he pressed for a wide variety of new starts with budgets that would quickly mount up in the years ahead to levels exceeding those of the Apollo era."[67] By April 1970 Paine, consistent with his vigorous and ambitious view of the post-Apollo era, decided NASA "would make the effort to fund something like a 200″ mirror in the southern hemisphere." This was to be "a first step in escalating our support to astronomy, and we picked the 200-inch telescope because it would be a very visible symbol of government continuing to support science." Meanwhile, NASA had "already augmented Brunk's program with an allocation for Aericibo [sic], and hope to get it up to $7–9 million in the next year."[68]

The Southern Hemisphere telescope never materialized. In fact, the severe budget cuts imposed on the overall NASA program after Apollo reduced the ambitious plans for solar system exploration as well as every other program in NASA. Figure 4.4 shows the overall NASA budget for 1961–76, broken down according to the Office of Manned Spaceflight and Office of Space Sciences. The space science budget actually gained as a percentage of manned space flight from 1969 to 1973, although inflation eroded the budget's buying power. The peak of the space science budget around 1973 largely represents the Viking Mars landing program.

Yet even in the face of such funding difficulties, the agency managed to send exploratory spacecraft to all of the naked-eye planets during the 1970s, and to land Voyager's successor, Viking, on Mars in 1976 (see appendix 2). Rather than increasing to $7–9 million, Brunk's budget, now an official line item and no longer lumped in with "supporting research and technology-science," included the $1 million per year mentioned to NSF earlier in December, just enough to cover the funding for one major instrument project. The telescope construction portion of the ground-based planetary astronomy program, unable to support facilities construction in the optical as well as other wavelengths, shifted emphasis after 1970 toward radio, radar, and infrared facilities. Even with this effort to make the most of available funds, the research support suffered. Whereas Brunk had had a hard time in the early years "finding people to give the money to," he now faced the opposite problem.[69]

The overall NASA budget-tightening of the early 1970s was aggravated by related developments. The Mansfield Amendment forced greater demonstration of mission relevance to basic research supported by the military, and military agencies began to divest themselves of certain facilities and research. The obvious place to seek support for such facilities was NASA and NSF.[70]

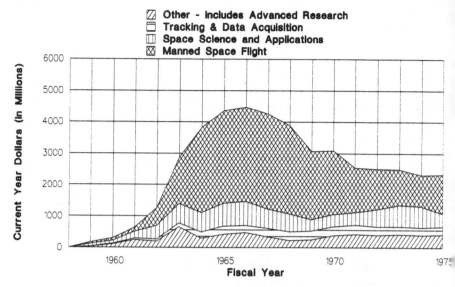

FIG. 4.4. NASA R&D Budget by Major Program Office. Source: Programmed amounts taken from tables 1-4, 2-4, 3-3, 4-4, and 6-2 in Vol. 3 of Linda Neuman Ezell, *NASA Historical Data Book* (Washington, D.C.: NASA, 1988).

One of these facilities on which NASA had relied was the 120-foot radome-covered fully steerable parabolic Haystack radio/radar facility built in MIT's Lincoln Laboratories for the Air Force in 1960. The Defense Department stopped its support of the Haystack facility in July 1970. NSF picked up the operational support of observations in the radio mode, and NASA had to pick up all support of the radar portion in support of the Apollo missions and planetary mapping. Haystack had been used for radar mapping of the moon, Mars, Venus, and Mercury, in refining solar system constants and distances, and in various relativity experiments. But basic operational support of the facility in the passive radio astronomy mode was about $1 million a year, with the active radar mode costing another half million dollars. As NASA's budgets tightened, assuming the operational support of Haystack would have had "disastrous" consequences on the planetary astronomy budget. A better alternative for planetary radar astronomy was the Arecibo Observatory.[71]

The Arecibo Observatory, completed in 1963 for study of the ionosphere, was another facility "abandoned on NSF's doorstep" by the military. Consisting of an antenna suspended by towers and wires over a thousand-foot natural valley and sinkhole, it was used for radio and radar astronomy at 50-cm wavelengths. NSF assumed responsibility for the facility in 1969 and renamed it the National Astronomy and Ionosphere Center. In June 1970, NSF obtained congressional approval to start a program to upgrade the surface of the antenna to be accurate to 6 cm, so that it could be used at those wavelengths, by replacing the existing chicken wire surface with aluminum panels. By 1974, the $5.7 million job was completed and the panels were being adjusted. In 1971 NASA and NSF concluded an agreement whereby NASA would provide $3 million for a high-powered S-band radar antenna and feed system to make the Arecibo facility the most powerful radar facility in the world.[72]

The final major facility for the 1970s constructed with NASA planetary astronomy program funds was the large infrared telescope recommended by the Hall Report and many other studies. In July 1969 Brunk held a meeting at JPL to discuss the future of ground-based infrared astronomy, and laid plans for an extensive survey to assess the quality of astronomical seeing at the site. He commissioned Caltech's James Westphal to develop, assemble, and calibrate the instruments and to analyze the data from the several sites. In March 1971, while the results of the 10-micron site survey were being analyzed, Brunk proposed the 3-meter infrared telescope as a fiscal year 1973 new start. Detailed design studies of the instrument were to begin during 1972, and after selection of the site and host institution further work could start. He then estimated the cost of the facility at $5 million, $2 million in each fiscal year 1973 and 1974, and $1 million in fiscal year 1974. It was to be in operation well before the 1977 Mariner Jupiter-Saturn mission (later re-named project Voyager, not to be confused with the Mars Voyager). The site survey recommended Mauna Kea, Hawaii, but a dispute broke out over the site selection. The instrument had severe develoment problems, difficulty getting through Congress and the Bureau of the Budget, and was finally dedicated in July 1979, the very month that Voyager II flew through the Jovian system and four months after Voyager I had passed through. Nonetheless, it proved to be an important addition to the world's stock of astronomical instruments, and worth waiting for.[73]

This was as close as the agency came to getting its ambitious major program of solar system exploration and space astronomy backed up by a correspondingly augmented ground-based astronomy program. Indeed, any Apollo-type goal was, in the budget austerity of the time, clearly out of the question. Paine continued his attempts to get as ambitious a program as possible, but resigned in September 1970. His influence did not reach far enough into other segments of the government, without whose coopera- tion he could not move forward with the program he desired.[74] Future NASA administrators had to reconcile themselves to tighter budgets and more modest ventures.[75]

In the planetary program, a balanced but much reduced effort was mounted. A Mariner gravity-assist mission traveled to Venus and Mercury in 1974, Pioneers were launched to Jupiter in 1972, and a Mariner (later "Voyager") Jupiter-Saturn with a Uranus option was scheduled in place of the more ambitious and costly Grand Tour. A Pioneer Venus mission consisting of one orbiter and a multiprobe bus was set for 1975.

The lack of "advance scouts" in the expanded but relatively modest planetary program meant that more would be required of ground-based planetary astronomy. But at the same time the budgetary constraints kept program funds approximately even after inflation. NASA was committed, in the early 1970s, to broadening the planetary facilities beyond optical into infrared, radio, and radar, as well as continuing support for the optical facilities already constructed, and maintaining the support to individual investigators for the actual research. Ground-based planetary astronomy's place and role in the planetary program, if not the level of support, was relatively secure.

Figure 4.5 shows the funding available from the NASA planetary astronomy program in three areas—research support to principal investigators at universities, research support to NASA centers, and major facilities construction. In the major facilities plot, two periods of active telescope development can be seen. The first, centered around 1966, represents the construction of the Texas and Hawaii telescopes, as well as support for smaller instruments during that period. The second, 1971–74, represents the upgrading of the Arecibo radio telescope for planetary radar. For comparison, figure 4.6 shows the amount NASA was spending just on astronomy research grants (excluding facilities) and the amount being spent by NSF on planetary astronomy is also shown. The specific dollar amounts are given in tables 4.3–4.5.

Originally, JPL and other astronomers had had difficulty convincing NASA that planetary astronomy had anything significant to offer, much less that NASA had a responsibility to support these ground-based studies and facilities. By the early 1960s the need for additional ground-based study had been demonstrated, and NASA moved to use its supporting research and technology funds to support individual investigators and some telescope construction. Construction and support of the two major optical instruments and the other facilities was spurred by the immediate needs of Voyager Mars, and even here the attitude persisted that this was a temporary mission need and that the ground-based techniques would rapidly be superseded by probe results. In anticipation of post-Apollo planetary exploration, plans were laid for an ambitious expansion of the ground-based and space-based planetary program into new types of facilities and study of planets other than Mars. While the level of support for this work was drastically curtailed relative to the ambitious plans, the spirit of the expansion was implemented.

Perhaps an unlimited number and variety of probes, orbiters, and landers to all of the planets, their satellites, the asteroids and comets would have made ground-based techniques obsolete. Instead the constrained planetary program had just enough probe missions to keep the ground-based program defensible to Congress and the Office of Management and Budget as a necessary adjunct to the spaceflight program, but not so many that the ground-based program would run out of work to do. A program with enough probe-based missions to make ground-based planetary astronomy "obsolete" would indeed have been staggering in its scope and cost.

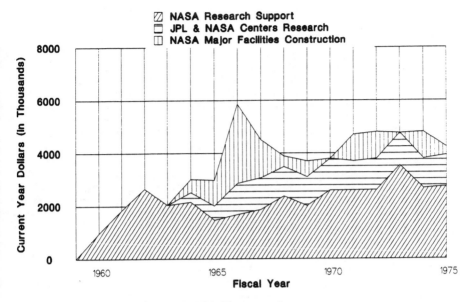

FIG. 4.5. NASA Planetary Astronomy.

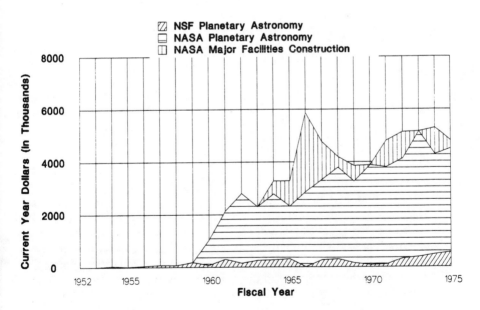

FIG. 4.6. NASA Planetary Astronomy—Including Facilities. (NSF figures provided for comparison)

TABLE 4.3

Distribution of NASA Funds for Ground-Based Planetary Astronomy (in Thousands).

Source: 1960–63 data compiled from ledger sheets in folder, "Funding-Past," William E. Brunk, AR; 1964–72 data from *Astronomy and Astrophysics for the 1970s* (Washington, DC: NAS, 1972), vol. 2, Reports of the Panels, table 9.28; data for 1973–75 compiled from looseleaf binder, "404s Written," William E. Brunk, AR. "-" = data not available or not clear. Major facilities column indicates year of initial contract. Texas 107-inch completed in 1969, Hawaii 88-inch completed in 1970. Planetary Patrol managed by Lowell Observatory. Arecibo upgrade included radar transmitter, fund transfer to NSF. IRTF = Infrared Telescope Facility; site survey and negotiations 1970–75, but no major funding.

Year (FY)	Major Facilities	Research Support	JPL & NASA Ctr	Major Facil Description
1959	0	—	—	
1960	0	976	—	
1961	0	1,835	—	
1962	0	2,647	—	Kuiper 60″
1963	—	2,041	—	
1964	500	2,154	355	Texas 107″
1965	975	1,467	535	Hawaii 88″
1966	3,000	1,679	1,180	Other facils
1967	1,447	1,866	1,195	
1968	400	2,371	1,118	Plan Patrol
1969	600	2,005	1,095	
1970	0	2,587	1,213	
1971	1,000	2,592	1,108	Arecibo
1972	1,000	2,605	1,195	Arecibo
1973	—	3,539	1,224	
1974	1,000	2,670	1,130	Arecibo
1975	300	2,736	1,200	Arecibo / IRTF

The irony is that at the very time that NASA top management had become enthusiastic about the ground-based astronomy program (planetary and other), and had taken steps to increase its support and to broaden the restrictive Dryden policy, political and economic developments had conspired to tie their hands. Seen from the standpoint of the massive Apollo funding and the equally ambitious planetary exploration plans for the post-Apollo era, the resulting program seems like a disappointment. However, from the standpoint of planetary astronomy in the 1940s or 1950s, and even the early 1960s, the resources available thanks to NASA (intellectual as well as financial and instrumental), were enormous.

TABLE 4.4
Total NASA Funding of Ground-Based Planetary Astronomy
(in Thousands)

Source: See table 4.3.

Year (FY)	Total Rsch/Ctrs	Total Incl Facils
1959	—	—
1960	976	976
1961	1,835	1,835
1962	2,647	2,647
1963	2,041	2,041
1964	2,509	3,009
1965	2,002	2,977
1966	2,859	5,859
1967	3,061	4,508
1968	3,489	3,889
1969	3,100	3,700
1970	3,800	3,800
1971	3,700	4,700
1972	3,800	4,800
1973	4,763	4,763
1974	3,800	4,800
1975	3,936	4,236

TABLE 4.5
NSF Research Grants in Astronomy, 1952–1975
(in Thousands of Dollars).

Source: NSF Annual Reports, 1952–75. "Astronomy" column taken from Astronomy Program headings. "Planetary" is taken in the widest sense, including any projects funded under the astronomy program which involved planetary research, optical and radio, even if such projects also included non-planetary work. 1973–75 planetary figures from Solar System Astronomy Program. "Percent Planetary" represents percentage of total astronomy dollars. We thank Lee Brakeiron for providing a computer listing of 1973–75 research grants funded under the Solar System Astronomy Program.

Year (FY)	Astronomy No.	$	Planetary No.	$	Percent Planet.	All Fields No.	$
1952	1	8	0	0	0	97	1,074
1953	7	81	2	9.7	12.0	173	1,698
1954	19	148	3	39.5	26.7	374	3,888
1955	19	363	3	14.5	4.0	588	7,857
1956	27	226	5	45.1	20.0	734	9,655
1957	33	454	7	79.8	17.6	997	15,529
1958	33	1,018	3	66.0	6.5	1,133	20,038
1959	51	1,955	7	184.0	9.4	1,809	49,122
1960	62	2,170	3	69.0	3.2	1,995	61,918
1961	54	2,151	4	297.2	13.8	2,102	69,036
1962	73	3,333	4	147.6	4.4	2,572	96,082
1963	69	3,702	7	254.1	6.9	2,714	117,213
1964	76	4,528	5	267.1	5.9	3,105	114,988
1965	74	4,520	5	307.5	6.8	3,228	122,238
1966	94	6,708	1	5.4	0.1	3,647	157,755
1967	118	5,900	5	267.0	4.5	3,972	172,600
1968	118	6,100	8	303.4	5.0	3,874	170,600
1969	125	6,820	5	156.3	2.3	4,088	176,020
1970	108	5,800	2	83.3	1.4	3,817	161,710
1971	135	6,420	2	99.5	1.5	4,329	174,560
1972	135	8,010	6	327.8	4.1	5,955	261,110
1973	170	8,800	7	393.2	4.5	6,138	268,050
1974	161	9,300	9	485.1	5.2	6,400	281,030
1975	188	10,930	12	566.8	5.2	7,019	338,000

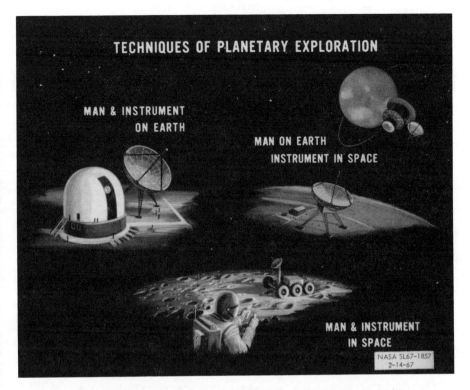

FIG. 4.7. By 1967 ground-based planetary astronomy had become fully included in NASA's roster of "Techniques of Planetary Exploration." Graphic courtesy of NASA.

FIVE

Conclusion

A Home for Planetary Science?

Previous chapters have traced, in narrative fashion, the varying state and status of planetary astronomy from the controversial work of Percival Lowell through the NASA years. The NASA ground-based program was designed to secure needed observations and analysis, as well as to help produce a new crop of planetary astronomers and planetary scientists generally. In this chapter I assess the purported impact of the NASA program on the various disciplines, understood as activities and as communities, that contributed to this interdisciplinary field. I also look at the preceding narrative from several theoretical perspectives of the social studies of science and technology.

IMPACT OF NASA FUNDING AND DEMANDS ON ASTRONOMY—THE RECEIVED WISDOM

A number of authors have discussed the impact of NASA's activities on astronomy. These works were written by, or relied heavily on accounts by, participants in planetary science and planetary astronomy. They should therefore be taken as expressions of the participants' self-image, against which the account in preceding chapters and other information discussed below can be compared.

A Columbia University study, commissioned by NASA in 1975, examined the impact of the agency on several different fields, one of which was astronomy. The authors of that study found that

> the demands of space technologies changed the very nature of large parts of astronomy, fractionating it into well-defined specialties even as it integrated other specialities into new groupings. Most strikingly it created—or is creating— out of a small, university-bound science, a Big Science, with all of the problems of manpower, organization, and public policy that such a development entails.[1]

As to the former assertion, the space-based results of various astronomical missions provided data impossible to gather from earth, and were bound to have an important effect on all portions of astronomy. The study of spectral regions hitherto unobservable because they are blocked by the absorption of the earth's atmosphere, such as X-rays, would be expected to give stimulus to a science whose observational basis had been constrained by a limited (ground-based) perspective.[2] Study of the planets had been constrained not by the absolute impossibility of observation but by the difficulty of conducting and interpreting such observations. Thus high-energy astronomy, whose wavelengths of light are visible only from above the atmosphere, had never really had a chance to begin before space techniques were developed, even though theoretical predictions had indicated certain kinds of high-energy observations that ought to be obtainable. Planetary astronomy, on the other hand, had been a vigorous specialty until the early part of this century. So high-energy astronomy was in a sense created by the availability of space techniques, while planetary astronomy was, according to the Columbia study, "resurrected."[3]

These new observations, far surpassing anything of which earth-based telescopes were capable, needed to be incorporated into a theoretical framework. Geophysicists, moreso than astronomers, had developed that framework, since it involved highly detailed studies of the planets such as geophysicists had made of the planet earth. In his historical memoir, *Beyond the Atmosphere,* Newell called particular attention to this circumstance:

> No longer was the study of the planets solely a venture of the astronomers. The dearth of new data that had led planetary studies into the doldrums and even disrepute among astronomers gave way to a sudden flood of new information that reawakened the astronomers' interest. . . . data on the moon and planets that poured in from astronauts and instrumented spacecraft . . . afforded earth scientists the opportunity to begin the serious development of a science of comparative planetology.[4]

Thus, according to Newell, the contribution of NASA's planetary exploration program involved the earth sciences as much as astronomy. Any specialty or discipline which had taken any any aspect of the physics of the earth as its topic could now be applied to other planets as well.

> But probably the most significant impact of space methods on geoscience was to exert a powerful integrating influence by breaking the field loose from a preoccupation with a single planet. . . . No longer restricted to only one body of the solar system, scientists could begin to develop comparative planetology. Insights acquired from years of terrestrial research could be brought to bear on the investigation of the moon and planets, while new insights acquired from the study of the other planets could be turned back on the earth.[5]

Moreover, as the preceding chapters show, the brief encounters with the planets and long periods of time in between such encounters often threw the follow-up observations back into the camp of earth-based astronomy. There thus developed a complicated and confusing interdisciplinary activity, often called planetary science(s), in which practitioners from many specialties and disciplines used widely varying methods and theoretical bases to study the planets. Earth-based observations using telescopes, which required astronomical training and involved workers in the astronomical community, gave rise to data that could only be fully understood by drawing on the intellectual resources of other specialties and disciplines.[6]

So Newell and the Columbia study identify a real pressure on the disciplinary structure of astronomy and geophysics brought by the new data which came in as a result of both the space-based and ground-based planetary program. But this gives too much credit to the techniques themselves. As the preceding narrative and the discussion below shows, the revision of boundaries in astronomical and geophysical sciences was as much a deliberate project conducted by scientists and administrators as an effect of data and methodology.

As to the Columbia study's assertion concerning the transformation from little to Big Science, one might object that in many ways astronomy had been a Big Science for many decades before the age of government largess, since astronomers always depended on special capital-intensive facilities, and some had even used an assembly line organization for large routine projects in the late nineteenth century. But a variety of independent evidence suggests that astronomers were among the slowest to join the postwar trend toward reliance on substantial government support.[7]

A number of more recent studies credit the planetary exploration program with renewing interest in ground-based planetary astronomy, and note the relatively poor knowledge of the planetary system before the advent of space techniques. A NASA advisory body called planetary astronomy a "previously neglected observational science . . . which has provided a large share of the discoveries about the solar system" made since the early missions.[8] A planetary science working group of the National Academy of Science's most recent Astronomy Survey Committee conceded that study of the planets had been "unshackled in the 1960s from the limitations of purely earth-based observations," but called attention to the "remarkable number of unexpected discoveries made through ground-based research" and found a "current vitality in this field that has not been widely appreciated even in the planetary science community itself."[9]

DISCIPLINE STUDIES—SCIENTIFIC JOURNALS AND SCIENTIFIC SOCIETIES

One of the indications that a fledgling scientific specialty has "come of age" is often the founding of a scientific society or a journal devoted to that specialty. One of the suggestions of the 1968 Hall Report on planetary astronomy was that "the establishment of a national society for planetary sciences or of an affiliate of an existing society would be highly desirable to serve as a forum for discussion and a cohesive force to facilitate recruitment of personnel, to assist in obtaining financial support or facilities for projects of unusual merit, and to encourage publication of results."[10] This recommendation followed a discussion of the fact that planetary study had no recognized "home" in the university curriculum, in part due to its interdisciplinary nature.

Attempts were made to establish interdisciplinary planetary science organizations under the umbrellas of two different scientific societies. The events surrounding these attempts provide an interesting view of the effects of the planetary program on scientific organization, as well as on the tactics of scientific organizers and promoters.

GEOPHYSICISTS

There had been an attempt to establish such a division within the American Geophysical Union (AGU) very early in the space program. The Executive Committee of the AGU had been discussing the problem of where space research should be located within the union, when Robert Jastrow and Gordon MacDonald wrote AGU president Lloyd Berkner proposing the creation of a section on "planetary physics." After discussions with various AGU members, the Executive Committee created an all-union Planning Committee on Planetary Science (PCPS), with Newell as chairman and Robert Jastrow as secretary. The PCPS had authority to arrange meetings, symposia, and periods for reading and discussion of papers at the AGU meetings.[11]

While the PCPS went about its task of arranging forums for presentation of papers on planetary science at the AGU meetings, the possibility of a formal separate Planetary Sciences Section for the AGU was raised and Newell was asked to prepare a report for the AGU Executive Council.[12]

In June 1960, Newell, Jastrow, and MacDonald drafted an essay, "A 'Home' for Planetary Sciences," for the AGU *Transactions*. While this essay stopped short of calling for a separate section, it laid the rhetorical groundwork for such action. First, the authors noted that with space techniques the time was coming when the planets could be studied using geophysical techniques, and pointed out the many ways in which the other planets were similar to the earth. "Thus, the techniques to be employed are those of the earth sciences, and not those of astronomy. The new areas of planetary exploration . . . will utilize many of the techniques already available in the earth sciences."[13] Thus they defined planetary science to be coextensive with the scientific purview of geophysics.

While the planets had been "not too highly valued a property" of the astronomers, those astronomers who were so inclined would continue to study the planets. But in addition, geophysicists "ought to join in." They saw a need for educational institutions to begin to educate workers to a broader framework, encompassing all of the planets in a single view. As to the proper home for planetary sciences, the astronomers no longer had exclusive claim—the interest of the American Astronomical Society was "on the basis of tradition and the fact that the planets are indeed astronomical bodies." The AGU, however, "has the strongest interest in this matter. For . . . the planets are indeed sisters to the earth in their relation to the solar system."[14] Thus they rhetorically suggested some competition with astronomy, that astronomy had given up the planets, and hence that the geophysicists had the opportunity to capture the prize, planetary science, which was the wave of future geophysics. After the spring AGU meeting, Berkner asked Newell to poll various members and prepare a report on the desirability of a separate section. Newell did so, and presented the report in March 1961.[15]

He began by reviewing the PCPS's achievements over the past year, in participating in arrangements for two national AGU meetings and a jointly sponsored AGU–American Physical Society symposium. He then presented several alternatives for planetary sciences within the AGU. (1) An entirely new scientific society devoted to space and planetary research could be formed, but this would have to draw on the overall community for financing and would compete with other organizations for funds, skills, personnel, and energies. (2) A joint commission among the various scientific societies with interests in this area could provide coordination and draw on a broad base of support but with the disadvantage that it would provide "no real home" unless the participating societies were to create similar units within their own organizations. (3) The present PCPS could be continued, but that would keep planetary sciences in their position of secondary status and power relative to existing sections. (4) A separate Planetary Sciences Section could be created wihtin the AGU with full status and power relative to other sections. This last alternative was clearly favored by the PCPS, and the committee contrived to make it the most attractive of the lot to the AGU members. To the argument that it would overlap existing sections, the PCPS replied that there was already overlap among other sections within the AGU. A "common denominator of research interest and training" was what unified existing sections and separated them from others. Such a common denominator was already possessed *de facto* by those using space techniques on planetary bodies, and such a body of research had already appeared in this field that "the existence of a professional planetary science group is now established as a manifest fact." It would be simple *de jure* recognition to formally establish a Planetary Sciences Section. The PCPS called for the AGU to establish the separate section, make overtures to the American Physical Society, the American Astronomical Society, and other societies to establish such sections within their own organizations, and establish a joint intersociety coordinating committee.[16]

These three motions were approved in Newell's poll and referred to the council for appropriate action. But due to a "rather deep disturbance on the part of some people concerning the problems involved," the only action taken was to appoint another committee, chaired by Philip Abelson, to study the question of a home for planetary sciences within the AGU, and how the specialty could be accommodated with minimal disruption of the organization.[17] MacDonald and Abelson, in consultation with other members of the Abelson committee, prepared a draft proposal for circulation and comment, suggesting an Interplanetary Sciences Section be created to deal primarily with particles and fields in space, allowing planetary topics to "inevitably find their home in the older sections of the Union."[18]

The Abelson committee submitted its report in July, recommending "with some divergence of opinion" that the AGU membership be asked to vote on the establishment of a new Section of Interplanetary Science and an intersectional Committee on the Study of Extra-Terrestrial Planets. "The committee recommends that there be no other immediate change in the structure and name of the Union. In making changes the costs, which in this case are considerable in money, prestige, and the values inherent in tradition, outweigh any visible benefits."[19]

As the council meeting approached at which action on the proposed changes would finally be taken, Newell, Jastrow, and the rest of the PCPS argued strongly that the new section should retain the name and focus of planetary sciences. The PCPS position was finally approved at the council meeting of April 25, 1962.[20] In November, after a mail ballot was passed by the AGU membership, a Planetary Sciences Section was created with Newell as president, Jastrow as vice president, and A. G. W. Cameron as secretary.[21]

The AGU Planetary Sciences Section debate shows how potentially disruptive the new space techniques could be to the established disciplinary structure of the sciences. The AGU members involved in the debates had all been a part of the International Geophysical Year, and most were closely involved in using space techniques. They were anything but scientific conservatives in their appreciation and acceptance of the new technologies for study in their particular fields. But when it came to finding a home within the structure of the AGU, which paralleled the disciplinary divisions found in university departments, the interdisciplinary diversity of planetary science, its unusually unified community of workers, and the use of space techniques raised serious problems. In the end, the creation of the Planetary Sciences Section was the result of strong lobbying and persuasive argument on the part of Newell, Jastrow, and others on the PCPS. While the argument proceeded, the PCPS took pains to demonstrate the vitality and *de facto* existence of the proposed section by setting up symposia and sessions of papers on planetary science at the various AGU meetings, and promoting papers on planetary topics at the individual section sessions at the meetings. Statistics on the number of papers presented at various meetings were frequently cited to show that planetary science was growing within the AGU and deserved a home.

The planetary sciences section continued to grow into the late 1960s, in part representing the intense work being done on the magnetic fields and particles of the earth, Mars, Venus, and interplanetary space, and the geophysical work being done on the moon. In the late 1960s the planetary sciences section of the AGU had grown to such an extent that, as part of a general reorganization of the structure of the AGU, it was split into planetology, which considered surafaces and interiors of planetary bodies, and solar-planetary relationships, which considered particles and fields and their interaction with the solar wind.[22]

ASTRONOMERS

Very shortly after the recommendations of the Hall panel on planetary astronomy in August 1968, the American Astronomical Society (AAS) approved the formation of a Division of Planetary Astronomy, with Joseph Chamberlain of the organizing committee. A preinaugural meeting was held at the University of Texas in December, to coincide with the entry into service of the NASA-sponsored 107-inch planetary telescope. By the spring of 1969 the first bylaws of the group, now named the AAS Division for Planetary Sciences (DPS), were adopted. The DPS was to exist

> for the purpose of advancing the investigation of the solar system with special encouragement of interdisciplinary cooperation. . . . Studies of the Earth and Sun fall within the scope of the Division's interests, insofar as they are oriented toward an understanding of the planetary system in general.[23]

Although originally the membership of the DPS was limited to AAS members (primarily astronomers), in March 1973 the bylaws were changed to allow a new category, affiliate membership, "open to planetary scientists who wish to be associated with the Division . . . but who are members of other professional organizations actively concerned with planetary science." At the same time, a publications committee appointed two years earlier under Clark Chapman recommended that the DPS endorse *Icarus* as the primary publication for planetary research, and that members attempt to publish in the most frequently read journals.[24]

This recommendation was the result of the committee's investigation of a severe problem in planetary science, the lack of journals devoted to the field. Using various survey analyses of publication habits and a poll of the membership, the committee found that at least sixty journals published planetary science articles, but aside from *The Moon, Meteoritics,* and *Icarus,* most devoted only around 10% of their pages to planetary topics. "In order to have access to as much as 90% of important articles, the library for a planetary science group must subscribe to 15 or 20 journals; however, over 90% of the pages of such journals contain articles irrelevant to planetary science."[25] The committee found six journals which were the most widely used and highly regarded by planetary scientists, which journals published eighty percent of the most frequently cited articles in the 1960s, but found various problems of editorial restrictions, refereeing policy, publication delays, and quality with all of them. They considered various courses of action, but finally decided at the March 1973 meeting to endorse *Icarus.*[26]

Icarus had been founded in 1962, under the editorship of Zdenek Kopal and A. G. Wilson. Carl Sagan joined the editorial staff in 1968, and became editor in 1969. Sagan began to solicit more articles in planetary astrophysics, in contrast to the previous concentration of celestial mechanics in the journal. Table 5.1 lists other more or less specialized lunar and planetary journals that were established in response to the increasing activity in planetary study.

Thus by the criteria of scientific societies and scientific journals, it would seem that planetary science had become an established disciplinary "entity" by the early 1970s.[27] As what some have called the Elizabethan Age of planetary science (1970–75) progressed there was even some competition between the AAS-DPS and the AGU for the attention of the small community. As table 5.2 shows, the DPS membership grew steadily. In 1976, associate members accounted for about fifty persons, meaning that more than 250 affiliated members of the American Astronomical Society had joined the division.[28]

The case of both the American Geophysical Union and the American Astronomical Society, as well as the campaign elaborated in chapters 2 through 4 to establish planetary astronomy as an essential part of NASA's program of planetary exploration, reveal something far different than scientific workers simply joining together out of similar intellectual interest. They reveal scientists as people, competing for attention, deploying remarkable rhetoric and persuasion, and vigorously engaging themselves in a social project that also happens to involve the intellectual activities commonly associated with doing science. Science, technology, micropolitics, macropolitics, all shade into one another. Certainly individuals may specialize in one or more of these activities, or engage in them at different times. But science gets done, and we learn about the solar system, only through the integrated sum of these activities. They were all doing science, all

TABLE 5.1
Scientific Journals for Lunar and Planetary Research.

Year	Title (Comments)
1947	*The Strolling Astronomer* (Amateur publication of the Association of Lunar and Planetary Observers [ALPO])
1959	*Planetary and Space Science*
1962	*Communications of the Lunar and Planetary Laboratory* (G. P. Kuiper, private publication)
1962	*Icarus: Journal of Solar System Research* (Z. Kopal and A. G. Wilson, eds.)
1966	*Earth and Planetary Science Letters*
1967	*Solar System Research* (U.S.S.R.)
1968	*Icarus* (Carl Sagan, ed., various editorial changes)
1969	*The Moon* (Z. Kopal, ed.; year of Apollo 11, first lunar landing and sample return)
1973	*Icarus* (Changes to monthly publication and endorsed by AAS/DPS)
1978	*The Moon and Planets* (Formerly, *The Moon*)

TABLE 5.2
AAS Division for Planetary Sciences Membership and Meeting Attendance, 1969–1976.

Source: Data compiled from various sources, including published meeting reports in *Icarus.*
*Effort made to reduce number of papers given at meetings due to crowded sessions.

Meeting		Attendance	Membership	Papers Given
1st	1/21/70	180	—	—
2nd	2/1–3/71	—	—	56
3rd	3/20–24/72	185	175	113
4th	3/20–23/73	—	—	135
5th	4/2–5/74	—	220	175
6th	2/17–21/75	230	250	126*
7th	4/1–3/76	250	300 +	150

the time. Planetary science and the planetary astronomy associated with it were deliberately constructed and actively maintained at every step. The planetary scientists in both the AGU and the AAS were intensely concerned to arrange mechanisms of effective communication and organization in order to sustain and facilitate their work.[29]

DISCIPLINE STUDIES—STATISTICS AND DEMOGRAPHICS

In addition to the development of specialized societies and publications, another indication of the vitality of a particular scientific specialty may be found in the number of publications in that specialty, relative to the overall discipline and relative to other specialities.[30]

A variety of statistical and survey data has been collected on astronomers and planetary scientists for various published studies and reports. In addition, I was fortunate to have access to a unique data base on biographical characteristics and publishing habits of American astronomers between 1950 and 1975, developed by Thomas F. Gieryn, which Gieryn and I augmented with selected information concerning NASA and NSF funding of planetary research.[31] These data provide some quite interesting information concerning astronomy's research directions, its growth, and its researchers themselves. But the interpretation of this data is not without hazard, due to various methodological problems. Hence, in most cases the interpretations which follow are conservative, and rely on fairly gross features that emerged from our analysis.[32]

Figure 5.1 shows the number of astronomers with at least one publication in any area of astronomy from 1951 to 1975. The dramatic effect of the increased attention devoted to astronomy after Sputnik in 1957 is evident, as is the saturation of the astronomy "market" in the mid-1970s.[33] The figure also puts the planetary specialty in the context of the overall growth of the astronomy discipline. Only during the 1972–1974 period did planetary astronomy's growth match that of astronomy as a whole. In fact, the number of people actively publishing in planetary subjects seems to have begun to accelerate at a rate nearly equal to that of the discipline as a whole in the early 1960s, only to falter after 1965, recovering somewhat at the beginning of the next decade.

Figure 5.2 compares planetary astronomy with other astronomical specialties. Of particular interest is stellar astronomy, since this specialty was the most popular one throughout the period. The number of planetary publishers stays at about half the number of stellar publishers (astrophysical theory is a separate specialty) until 1959, at which time the planetary specialty accelerates more rapidly than stellar. After both take a slight dip in 1962, planetary becomes the equal of stellar in 1965, staying at that plateau until 1968–69 (when major NASA planetary telescopes started operation, a major planetary exploration program for the '70s was seriously being advocated, Mariners 6 and 7 flew by Mars, and the outer planets missions were being planned), when it again accelerates. The activity in planetary astronomy then lags behind stellar once again, before accelerating from 1972 to 1974. The peak in 1974 probably represents probe-stimulated activity, as Mariner 9 orbited Mars for a year, Mariner 10 flew by both

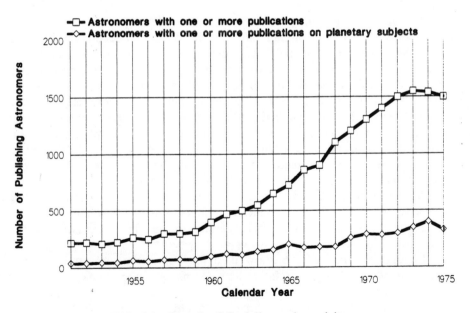

FIG. 5.1. Growth of discipline and specialty.

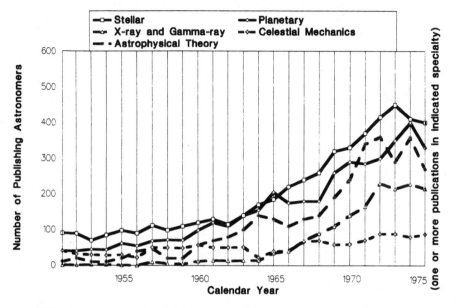

FIG. 5.2. Growth of specialties (one or more publications in indicated specialty).

Venus and Mercury, and Pioneers 10 and 11 flew by Jupiter. Planetary astronomy thus shows enough gain relative to other specialties to indicate some increasing interest, but hardly enough to be called a "resurrection." Important developments were occurring in other specialities, particularly in stellar and high-energy astronomy, which attracted new as well as established astronomers. The dramatic increase in celestial mechanics just after Sputnik represents the challenging new problems posed by satellites orbiting close to their primaries. The subsequent decline should perhaps not be interpreted as an absolute decline in activity in the specialty. As the space program developed, problems of celestial mechanics moved increasingly to engineering schools and curricula as "astro-dynamics"; much of the work in this field was published in journals not likely to appear in the index searched by Gieryn to build the publications database and was done by engineers not likely to hold membership in the American Astronomical Society.

Figures 5.3 through 5.6 show data for publications about individual planets, and the strong correlation between planetary missions and number of publishers. There is a clear increase in publishing activity on individual planetary problems around the time of planetary missions, but the activity drops off in most cases rather rapidly. The activity before the first arrival of a planetary probe, and to a lesser extent that between probe missions, is entirely ground-based. The dramatic spikes associated with probe missions perhaps accounts for the "noisy" curves of the planetary specialty. The activity among astronomers in planetary topics generally is somewhat less dramatic than some of the anecdotal testimony (Newell and the Columbia study, for example) would lead one to expect. At the same time the activity in planetary topics is far more intense than other survey evidence, reported below, suggests.

Other surveys conducted around 1970 found astronomers spending only 4–6% of their time on optical and radio observations of the planets, up to 39% of their time on objects beyond the solar system, and 44% on theoretical astrophysics.[34] A survey of astronomers' research interests from the same period found similar percentages.[35] In 1966–67 another survey found similar percentages for the overall use of major telescopes for lunar and planetary research.[36] All agree generally with the level of support provided by NSF for lunar and planetary research from its astronomy program (see chapter 4). They also suggest a basis for what some planetary astronomers have called the "ten percent rule," referring to the observing time and funding made available for planetary research by the astronomical community as a whole. The survey results discussed above suggest continuing lack of interest among the majority of astronomers at least to about 1970. A similar survey conducted around 1980 found little change.[37] These surveys also suggest that relatively few new astronomy Ph.D.s were taking up planetary research.

A survey conducted for the 1968 Hall Report on planetary astronomy was able to identify only eighty planetary science Ph.D. theses between 1960 and 1967. In commenting on this figure, the authors of the report noted "an insignificant fraction of NASA traineeships was used to support planetary-science training. It is difficult to understand why the annual rate is so low when the support available (until recently) for graudate study and research in planetary science has been both ample and diversified, and a national program of lunar and planetary probes was in operation."[38] They suggested that a "dominant reason for this may well be the fact that very few academic institutions have departments or other units committed to planetary science," and called attention to the fact that in three out of four cases planetary science was only a sideline in departments committed to other interests. They added that "the rapid development of new techniques has served to shift planetary research from the astronomy discipline to other departments," increasing the number of workers in the field and the number of Ph.D.s awarded, but questioned whether "they have always provided the best training for planetary astronomy."[39]

The results of that survey are reproduced here as figures 5.7 and 5.8. The authors noted that the production of planetary science Ph.D.s was not heavily weighted toward any single university nor any single department, implying that there was no real institutional "home" for the field (figure 5.7). Note also that in their graph of the number of planetary science

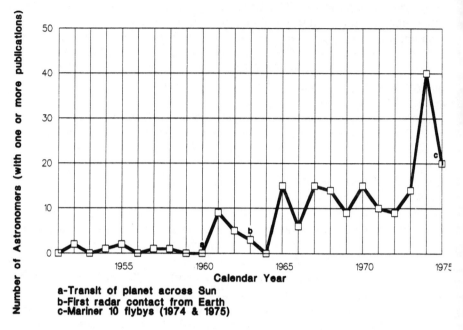

FIG. 5.3. Studies of Mercury.

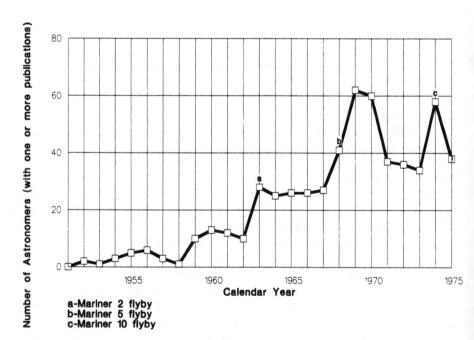

FIG. 5.4. Studies of Venus.

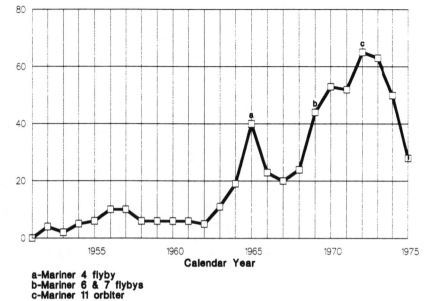

a-Mariner 4 flyby
b-Mariner 6 & 7 flybys
c-Mariner 11 orbiter

FIG. 5.5. Studies of Mars.

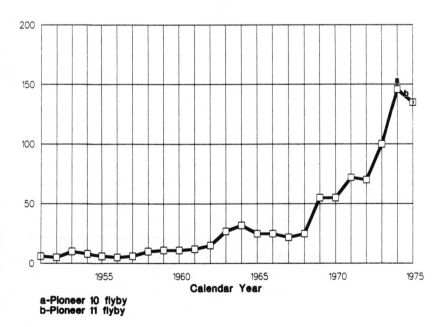

a-Pioneer 10 flyby
b-Pioneer 11 flyby

FIG. 5.6. Studies of Jupiter and Saturn.

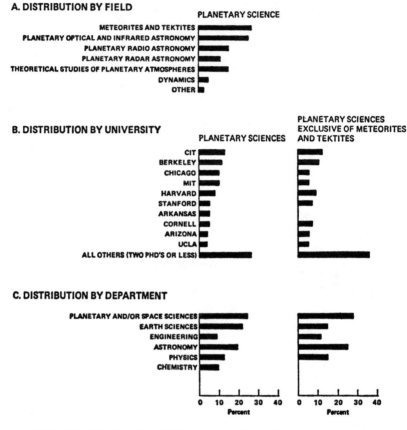

FIG. 5.7. Distribution of Planetary Science Ph.D.s, 1960–1967. Source: Panel on Planetary Astronomy, NAS, *Planetary Astronomy: An Appraisal of Ground-Based Opportunities* (Washington, D.C. NAS Pub. no. 1688, 1968), p. 67. Planetary science does not include space physics in this breakdown.

Ph.D.s granted between 1960 and 1967 (figure 5.8), the rate is greatest between 1961 and 1965, after which it falters significantly. This, curiously, is a persistent feature of the statistical data on publications as well.

The apparently low number of NASA University Program predoctoral trainees in planetary astronomy may be related to the purported lack of identification with NASA and the space program which one study found among NASA trainees. NASA gave the university up to three years of support for each trainee, but the university, not NASA, selected the recipients. Some trainees were unaware their funding was coming from NASA.[40] Thus if planetary research were not firmly established as a research area

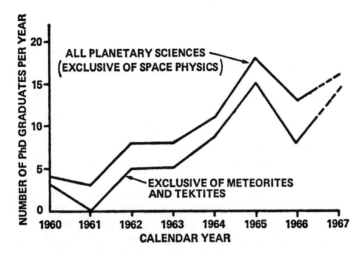

FIG. 5.8. Planetary Science Ph.D.s Granted, 1960–
1967. Source: Panel on Planetary Astronomy, NAS,
*Planetary Astronomy: An Appraisal of Ground-Based
Opportunities* (Washington, D.C.: NAS Pub. no.
1688, 1968), p. 66. The number of Ph.D.s granted
in physics, astronomy, and all earth sciences in 1965
was about 1,000, 65, and 400, respectively.

in academic science, or had no real institutional "home," then support
dispensed by the universities would not likely find its way to planetary
sciences.[41]

During the period from around 1965 to 1975 the entire discipline of
astronomy was growing at a phenomenal rate, and various other specialties
within astronomy were perhaps more atractive to new Ph.D.s. Stellar
astronomy had its share of exciting discoveries during this period, and
attracted many young researchers. High-energy astronomy, especially, un-
derwent a period of dynamic growth at the same time.[42]

Apparently, NASA found an older core of astronomers who were willing
to shift from other research areas to the planetary specialty (or to undertake
planetary research concurrent with other work).[43] Around half of all planetary publishers stayed exclusively within that specialty, with the rest maintaining a broader interest.[44]

As to NASA's direct responsiblity for the existence of a planetary science
and planetary astronomy community there can be little doubt. Of those
publishing just one or more articles in a given year, an average of 15%
through the 1960s, increasing to more than 20% by 1975, were Principal
Investigators on NASA research grants. Of the more prolific publishers,
an average of about 25% through the 1960s, increasing to 35% by 1975,
were Principal Investigators. The money made available just for research

grants extended far beyond the official investigator of the project, and the research facilities not included in these analyses must be counted as additional NASA support extending throughout the planetary community.[45]

The sketch of the planetary science community (most of whose members were affiliated with the American Astronomical Society, and thus had strong astronomical interests and ties) which emerges from these statistics suggests a small community, growing from around 100 members in 1960 to a peak of about 400 members around 1974, but at a rate substantially less than that of the astronomical community as a whole. Around half worked exclusively on planetary topics, the rest maintaining interest in other astronomical specialties. Thus there was a core of around 200–250 planetary workers allied with the AAS.[46]

But from around 1965 those publishing exclusively on planetary topics represented a decreasing percentage. The production of new Ph.D.s in planetary science appears to have remained relatively stagnant during the 1970s, at around 20 per year, and the data suggest that planetary specialists in the AAS were generally somewhat older than other specialists.[47]

DEPENDENCY OF PLANETARY ASTRONOMY ON NASA

A congressional Office of Technology Assessment report on the state of the space sciences made a special point of the dependency of certain scientific specialties and disciplines on government support. The authors noted that in these cases "there is a kind of social contract between scientists and society, in which the pursuit of knowledge is exchanged for economic support."[48] The increasingly detailed knowledge which follows large government efforts in Big Science often results in the founding of additional specialties and subdisciplines, "and each of them requires continued public support if it is to advance further."[49]

In less than two decades astronomy and planetary sciences had experienced wide fluctuations in funding, employment, and general interest, a pattern familiar in the basic sciences in general. But in the case of the study of the solar system, which requires research tools far beyond the means of any but government groups, such trends have implications not just for the level of activity, but perhaps for the very survival of the science itself. In recent years, a number of articles and studies have appeared suggesting that in the absence of continuity of funding, some planetary science research teams were disbanding as their members struck out into new and more secure directions of research.[50] This recalls the comment in the 1968 Hall Report on Planetary Astronomy that "the absence of a greater number of active, specialized planetary science departments may reflect some skepticism about the permanence of the space effort." The authors predicted that if the federal government were to commit the nation "to a long-term planetary-exploration program, the rate of Ph.D. production in planetary science will probably increase as university confidence in the stability and continuity of the national program develops."[51]

Resources targeted to specific areas of potential research may attract more workers to such areas, but the decision to begin a major new line of research is seldom made lightly, and the decision to specialize exclusively in that line of research is even more rare.[52] An important consideration is always an estimate of the potential for results, as well as an estimate of the stability of a particular line of research. Stability and continuity become more important the greater the investment in training required. The real but somewhat muted renewal of planetary interest found in the publishing habits of astronomers may reflect these concerns. The fact that so many of the planetary astronomers maintained other (and more "mainstream") interests may also be a related phenomenon.

Planetary research was a scientific enterprise whose pace and direction of development ultimately depended on considerations far beyond the scientific desiderata developed by the research community itself. The clash of the goals and priorities of the research community with those of their sponsors (NASA, Congress, the public) has been described in several places in this study. One such clash, for illustration, was over the early emphasis on Mars and the use of large, complex, Voyager-type systems. Most of the scientific community came to favor a more balanced approach, using simpler spacecraft to visit several planets. But the engineering community favored the more challenging Voyager-class missions. In addition, NASA was concerned about whether the less glamorous agenda could capture the attention and enthusiasm of Congress and the public, and hence ultimately funding and project approval. While a detailed comparison of the atmospheres of Mars, Venus, and the earth held great intellectual interest for the planetary science community, the search for life on Mars with an advance robot explorer and the prospect of human footsteps on the Red Planet held far more allure for those who ultimately controlled the purse strings.

Richard Hirsh's history of X-ray astronomy illustrated, in part, "how the fortunes of the specialty have fluctuated as a consequence of its absolute control by political forces."[53] This is certainly true of planetary *science,* since it depended on the planetary spaceflight program. However, even though planetary astronomy did not depend so much on constant technical innovation, but rather on access to existing observational instruments, it was no less controlled by similar forces. Access to telescopes was essentially secured by NASA as a supporting requirement for the planetary flight program. As the degree of interest in the planetary flight program went, so went ground-based planetary astronomy, since in the absence of NASA research grants and operational support for the NASA-constructed telescopes, ground-based planetary astronomy could not be carried out. NSF did not move to increase its funding of planetary astronomy when the magnitude of NASA support faltered in the early 1970s, and the planetary programs on NASA-supported telescopes were continually in danger of being swamped by stellar and galactic programs.[54]

POLICY—LEVERS OF CONTROL AND
GOVERNMENT-SUPPORTED SCIENCE

One result of a scientific endeavor's dependence on government support is that some of the new specialities are in a sense artificial products, created and maintained at least in part by forces operating outside the science in question. The relative lack of attention given to planetary astronomy before NASA (and even during the NASA years by such organizations as NSF) can be seen as an assessment by the astronomical community of what constituted scientifically important and technically feasible research problems worthy of being attacked with the limited resources available. NASA's attempts to stimulate this field, while partially motivated by scientific sincerity and a somewhat broader vision concerning the scientific potential of space techniques, represented in some sense interference, an attempt at political control. One study of science policy suggested that types of government control deriving from an interest in solving specific problems "are focussed on a specific product value of science. They imply strong interference with the internal structure of scientific processes, because they presuppose that the internal rules which determine problem selection and the orientation of intellectual development can be influenced extensively by political guidance."[55]

Although NASA's large-scale entrance into ground-based planetary astronomy was oriented initially toward such a product—the atmospheric pressure on Mars—it was never totally oriented to so narrow a goal. Indeed, Brunk, Newell, and others explicitly rejected such an approach, although they did emphasize the product value in certain forums, such as congressional presentations and interactions with administrators and engineers. Direct but limited access to the planets required an active and scientifically vigorous community of planetary researchers able to integrate ground- and space-based results and bring them to bear on important theoretical problems. In order to do this, as well as support the probe program, access to ground-based instruments was required and the cooperation and support of the wider scientific community was no less important. NASA was led to establish the ground-based program most immediately because it was deemed essential to the engineering success of the planetary exploration program. But equally motivating for Newell at least, was the desire to produce a scientifically sound and productive program. But the scientific community, and astronomers in particular, did not leap at this new opportunity. Other factors were at work.

Wolfgang van den Daele and Peter Weingart contend that, if the desired research cannot be stimulated through ordinary scientific channels, the government "must supplement and reconstruct this system by social planning, by establishing institutes, creating careers, and setting up training courses."[56] The debate during the early 1960s within NASA over just how to go about getting the planetary research it needed centered on the question of how completely the agency ought to reproduce the scientific system in order to direct its emphasis. The proposed solution of the Air Force, to create a fully government-owned and -operated planetary facility at Cloudcroft, represents one extreme. To some extent Kuiper's Lunar and Planetary Laboratory and the JPL proposals also represented attempts to duplicate the system, and ensure that planetary astronomy and planetary research would have all of the resources it needed without having to appeal to the shared facilities of the astronomical community. NASA was unwilling and, within the constraints of the planetary program unable, to make such a commitment on such a scale.

The method chosen by NASA can be seen as a compromise: an attempt at more indirect control, through financial support of planetary work, conferences and symposia to drum up interest in planetary research, construction of facilities dedicated in part to planetary astronomy and other planetary research, and gifts to the astronomical community such as the facilities, interdisciplinary grants, and predoctoral training grants of the University Program. Behind this lay the hope that with such "seed" programs, and with expected successful results of the ground- and space-based programs, the overall community would respond with increased enthusiasm for planetary research. To assure access to the NASA-provided facilities while waiting for the longer-term stimuli to take effect, the contracts for such telescopes had "teeth" in them, language explicitly stating to what extent planetary researchers were to have access.[57]

NASA's use of these inducements seems to have been only partially successful. It was motivated in part by a desire to disrupt the usual practice of science as little as possible by avoiding the establishment of competing institutions which would then be abandoned after their mission relevance had vanished. After planetary exploration became more commonplace, NASA planners and scientists hoped, planetary science would naturally develop into a respectable scientific specialty or discipline of its own, and with or without planetary astronomy—depending on whether further uses for planetary astronomy would emerge—the new discipline would take its place within the economy of science. But NASA became identified as the sponsor of planetary astronomy, and the community looked to this agency to take care of it. After ten years of the expanded planetary astronomy program, Brunk commented that at certain institutions, "since NASA provides significant support for planetary research, each institution has put a larger fraction of its own money to staff its stellar program rather than its planetary program. This approach is unfortunate but understandable. They would be very happy to increase the size of the planetary staff *if increased funding were available* (emphasis mine)."[58] NASA had hoped to push the fledgling out of the nest, but it seemed unable to survive on its own. Several studies of other areas of astronomy during this time suggest that additional factors were at work.

COGNITIVE VS. INSTITUTIONAL FACTORS

With planetary exploration and planetary astronomy programs in place, planetary studies seemed ripe for advance. They had been primarily constrained not by theory but by fundamental data. Michael Belton described his field of planetary science as a "discovery-bound" specialty, and noted that discoveries were just as likely to come from ground-based or earth-orbital observations as from probes.[59] A working group of the Astronomy Survey Committee described planetary research during the 1970s in the following terms: "The character of research in planetary science during the past decade has been dominated by this acquisition of fundamental data, discovery, and exploration. Rationalization, interpretation, and explanation tend to be cautious and tentative. Questions proliferate. The solar system that is being revealed is one of contrast and individuality. Unexpected and strange phenomena abound."[60]

Thus cognitive factors would seem to suggest that astronomers would have jumped at the opportunity to work in the area. A data-starved astronomical specialty suddenly had markedly increased opportunities for both ground- and space-based observations, and an agency willing to fund such work. There was no shortage of interesting problems about the planetary system in the early 1960s, and they continued to emerge as various observations allowed the solution of some but opened up many others. Why then does there seem to have been such a muted response among astronomers to planetary studies and its opportunities?

Van den Daele and Weingart cite several examples to show that "institutional conditions may operate as factors of resistance irrespective of the cognitive stage of development of a discipline," and that "strategic variation of institutional factors designed to initiate or control scientific development may be frustrated by cognitive conditions." They conclude that "science policy has to rely in every case on an assessment of both the cognitive and the institutional conditions of science relative to the objectives of political control."[61]

Certain institutional factors such as the lack of an identifiable "home" for planetary science within the university and general academic structure, oversubscription at observatories, other demands on astronomy during the period, the general lingering perception of planetary work as Lowellian and amateurish, and the poor scientific reputation of NASA may have frustrated the cognitive conditions favorable for growth. Richard Berendzen got an interesting response from astronomers in a survey sent to every member of the AAS in late 1966. About two-thirds replied. Only 38% were in favor of teaching undergraduates about NASA's activities, and 24% "stated explicitly that only the results should be given." He received "virulent" comments regarding Project Apollo, and "about half the astronomers opposed teaching about Apollo while only a quarter supported it; roughly one-fifth said that only the scientific results, if any, accrued from Apollo should be taught." Apollo had come to symbolize NASA's approach to "science" for many scientists, and aggravated the stormy relationship between the agency and the scientific community. Berendzen noted that the NASA questions "evoked the greatest emotionalism of the questionnaire. For unspecified reasons some scientists harbored intensely derogatory feelings about NASA."[62]

Unfortunately for NASA, just as the agency was attempting in a serious way to mend fences with the scientific community (1968–70) and put scientific work into a more prominent position relative to the engineering aspects of space research, the budget was being slashed and various programs curtailed. Not only was the research of the various spaceflight program offices cut back, but the separate and important NASA University Program was eliminated entirely by 1970. The seemingly chaotic and ever-changing particulars of the planetary program during this period, much of which was reported in *Science,* did not send a message of confidence and stability to those contemplating engaging in planetary research.

POTENTIAL RECRUITS

Nigel Gilbert, in his study of the growth of radar meteor research, noted that the growth of a new research specialty "requires . . . that recruits be available to join the new research area, and growth will continue only so long as a continuing supply of new recruits is forthcoming. Eventually . . . the area may either decline in importance, gradually losing members to more interesting fields, or it may become closely associated with other 'neighboring' research areas."[63] There were certainly plenty of potential recruits available during the late 1960s and early 1970s, as the expanded university curricula drew large numbers of astronomy students. But as David Edge and Michael Mulkay note, the "recruitment of graduates and other researchers is often difficult in the period before major discoveries have been made and before the scientific reputation of the new specialty has been established. During this period students seldom choose to enter the new field unless they are actively encouraged by those already involved."[64] Few of the proponents of planetary research in the early years were located at teaching institutions. Most of the institutions where planetary science took hold were those where people such as Kuiper and Urey, or their students, were located.[65]

During the late 1960s, when most of the potential recruits were choosing areas in which to specialize, the two major NASA telescopes were just coming into operation, and the two chief problems of planetary astronomy, the composition of the Mars and Venus atmospheres, continued to bedevil researchers. The planetary astronomy being done during this period was of the most basic data-gathering sort. In short, planetary astronomy was not far removed from its observational stage of earlier in the century when interest in the planets among astronomers had waned. While the combination of planetary astronomy and planetary exploration had by the early 1970s revealed important information concerning the atmospheres of Venus and Mars and had dramatically altered conceptions of the Martian surface, it had by no means arrived at an unambiguous interpretation of the results. As Edge put it, "In the early stages of an innovation, it cannot be clear whether the new method is going to assist the established workers, or turn out to be a cuckoo in the nest. So support will only emerge if the newcomers are not seen as in competition for scarce resources with the established discipline."[66] In the case of planetary exploration, the planetary and astronomical communities were in competition for scarce resources on two fronts. First, the planetary exploration program was up against a large orbital astronomy program in the NASA budget, and NASA decided to emphasize planetary exploration and defer a large program of space astronomy. Second, the ground-based planetary astronomers were in direct competition with other astronomers for the most precious resource of all—telescope time.

Scientific disciplines can certainly tolerate and even enthusiatically support various specialties sharing common research facilities. But in the case of astronomy, these facilities were heavily oversubscribed. The entire system of astronomy—research instruments and facilities, graduate schools, personnel—was strained. This led to various specialties—planetary, stellar, galactic, etc.—with quite different goals and agendas appearing as competitors before the observatory visitor committee to compete for telescope time. As the planetary exploration program was cut back again and again through the mid-1970s, more planetary scientists turned to ground-based planetary astronomy as the only way to obtain new data to complement that acquired from the spaceflight program. For access to the required facilities, planetary as well as stellar and galactic astronomy had to be judged by common cognitive criteria—how would the proposed research program help advance the broad goals of astronomy as a whole?

In addition, there was no strong guarantee, especially in the early years, that planetary exploration would not turn out to be a "cuckoo in the nest." As for planetary astronomy, many astronmers thought that this specialty had long been shown to be just such a bird. Edge and Mulkay note that relationships between radio and optical astronomy groups "have been intimately bound up with the exchange of scientific information." They further assert that "in Britain, at least, the social and intellectual integration of optical and radio astronomy has proceeded smoothly, in parallel with the gradual increase in the perceived astronomical value of the information furnished by the newcomers."[67] In a similar way, high-energy astronomy, while totally different in technique from optical astronomy, nonetheless contributed valuable information which bore directly on problems of astrophysics and cosmology.

Of what astronomical value, then, was the information which could result from the research programs of planetary astronomers? How would an understanding of the composition and other characteristics of the moon and planets and other solar system bodies help astrophysicists understand the evolution of stars, or help galactic astronomers understand the broader concerns of cosmology? Ultimately, of course, the composition and evolution of the planetary system is bound up with the evolution of stars. The planetary system and its history can tell us much about the early history of the solar nebula and its differentiation. But it is a long, difficult route connecting planetary research with stellar research. Cosmogony meets cosmology only on the very fringes of each. The immediate products probably did not seem all that relevant to the main concerns of astronomy.[68]

Radio astronomers initially prosecuted their research with techniques and facilities quite separate from those of optical astronomers. Initially they were outsiders, but not competitors. When radio astronomers began to produce information which bore on astrophysical problems, then began the gradual integration of the two techniques into overall astronomy.[69] Planetary scientists pursued their research using separate facilities (probes). But planetary astronomers were in direct competition with the rest of the astronomical community for scarce telescope and instrument time, and the results of their research did not bear directly on the chief day-to-day problems of stellar and galactic astronomy. Moreover, while the techniques of planetary astronomy (photometry, spectroscopy, etc.) were superficially similar to those of stellar and galactic astronomy, there were important special twists to their application in the planetary area. Many stellar and galactic astronomers thought they understood the techniques the planetary astronomers were applying, and that it had all been done before with little concrete result.[70]

Edge and Mulkay spell out several conditions which, when present in a scientific situation, are correlated with less resistance to innovative research programs: "[Resistance] will be less evident, we have proposed, to the extent that innovators and existing practitioners make use of a common body of scientific knowledge, to the extent that new departures are seen as supplementary rather than incompatible with current views, to the extent that the innovators are regarded as having a special scientific or technical competence, and to the extent that the newcomers supply information that those already in the field find valuable within their own frame of reference."[71] The interaction between planetary astronomers and the rest of the community essentially failed all of these conditions. The information which the planetary astronomers were producing was relevant to planetary scientists (themselves on the fringes of several earth science disciplines as well as astronomy) and spacecraft designers, but not, in general, to other astronomers. What planetary astronomers shared most with other astronomers was their educational background, telescopes and instrumentation, and to some extent the techniques that they employed.

DIFFUSE IDENTITY

Planetary astronomy was not so much "resurrected" as it was transformed by space technology and the scientific, technical, social, and political milieu in which that technology was used to explore the planets. In place of an exclusively astronomical specialty there appeared a new heterogeneous community of researchers attempting to establish a professional identity. The planetary science which resulted could probably claim a closer contemporary theoretical kinship to the earth sciences than to astronomy. But in the course of the creation of this planetary science during the 1960s and 1970s, an important niche was found for planetary astronomy.

The small community of planetary astronomers which evolved to populate that niche had strong ties to the astronomical community by virtue of tradition, education, research tools, and the place of the solar system within the astronomical universe. But it was equally tied to the diverse planetary science community by virtue of its goal of understanding the various bodies of the solar system as planets. This diffusion of identity within the disciplinary structure of the sciences made it difficult for planetary workers and especially planetary astronomers to find a true "home" within the economy of science.

To a certain extent NASA became a surrogate institutional focus of planetary research. Inasmuch as NASA was almost the sole sponsor, such work could hardly have been carried out without the agency. But NASA's role went beyond the vast material support it provided. It also provided mechanisms for communication among the diverse workers, informal ones as well as the more formal symposia and conferences. The agency also contributed to the direction of the specialty's research, by selecting which planets were to be targets of the flight program and hence in which particular areas research would be supported.[72]

CONCLUSION

Astronomers, rocket pioneers, poets, and novelists had long dreamed of exploring the other planets. On "public nights" at observatories, astronomers forsake the galaxies and stars that ordinarily command their professional attention and turn the telescopes instead to whatever planet happens to be available, preferably Saturn. They know what the public wants. Similarly, when the Hubble Space Telescope—designed first and foremost for stellar and galactic astronomy—experiences first light, it will undoubtedly turn first to planetary targets for "public affairs images." There is nothing wrong with this. Planetary scientists before public audiences do the same— they forsake the esoteric data that is for them scientifically exciting and show pictures of Saturn. Implicit in these choices is the realization that there are more than scientific motivations for doing science, and that when the public foots the bill it deserves at least some return.

The early chapters of this history showed that many people had many different reasons for wanting to go to the planets: technological challenge, national prestige, lucrative business contracts, military dominance, scientific agendas, and not least the irreducible wonder and allure of exploring other worlds. Succeeding chapters recounted how impassioned and dedicated people assembled and deployed the many and varied resources necessary to accomplish what so many had dreamed of doing, and to accomplish their own personal and professional goals as well. There was no single overriding motivation—not science, not discovery, not ambition. In a fluid and ever-changing system of give and take, each person and group involved was at various times both exploiter and exploited, actor as well as resource for another actor.[73] As Michel Callon has written, "Right from the start, technical, scientific, social, economic, or political considerations have been inextricably bound up into an organic whole. Such heterogeneity and complexity . . . are present from the beginning."[74]

What Callon calls an "actor network" and Thomas Hughes calls a "system" always tends toward dissolution because the various heterogeneous components are not compelled to retain their assigned roles.[75] The planetary exploration system—in which ground-based telescopes, space probes, astronomers, engineers, accountants, politicians, and a host of other entities worked together to explore the solar system—required "system builders" such as William Brunk to sustain the many relationships. Using rhetorical persuasion, he did this by telephone, letter, and personal visits.[76] He did this by interacting with other astronomers, with other managers in the hallways of NASA, with his counterparts at other agencies, contractors building telescopes and parts of telescopes, and through a bureaucratic process with politicians at various levels of the government. He and others, as Hughes puts it, "were no respecters of knowledge categories or professional boundaries."[77]

The telescopes were themselves part of another system, traditional astronomy, and tended always to revert to their assigned role (studying stars and galaxies) within that system. Embodied in the very technology of the ground-based telescopes that NASA created and modified was a recognition of this. They were crafted—physically and in their administrative and fiscal arrangements as well—to do planetary as well as stellar astronomy.[78] The spacecraft were themselves part of a technology system and tended always to revert to their assigned roles (engineering virtuosity) within that system. Politicians making funding decisions from the standpoint of yet another system had to be educated continually as to why the space agency (NASA) was supporting work on the ground that was already the job of another agency (NSF).

In 1969 NASA administrator Paine attempted to incorporate the planetary astronomy–planetary exploration system into a much larger system aimed at sending people to Mars. These plans collapsed because he could not extend the system far enough. He was able to assemble such a system or network within NASA (and even there it was unstable) but not beyond. The other parts of the government were themselves woven into other, more coherent and stronger systems involving a host of domestic and international issues. The planetary astronomy–planetary exploration system then adjusted to a more modest level, where it achieved a relative equilibrium and stability during the 1970s. Even this stability began to dissolve late in the decade, as I suggested in the introduction, and has continued in this tenuous state to this day.

This is part of the story of how humans took the first steps to explore the neighboring planets. The events I have recounted should remind us that high technology and sophisticated science, in space as on earth, are intensely social and human—all too human—activities. These aspects of the story can help us appreciate and marvel at the scientific and technical achievements all the more. In the oft-quoted and perhaps apocryphal words of Wernher Von Braun, "We can lick gravity, but sometimes the paperwork is overwhelming."[79]

Appendixes

APPENDIX 1: A NOTE ON SOURCES AND ABBREVIATIONS USED

References to unpublished materials make use of the following abbreviations and conventions:

NHO, NHOA
: NASA History Office, NASA History Office Archival Document Collection, National Aeronautics and Space Administration Headquarters, Washington, D.C.

WNRC 255-79-0649: 25 (35) "Chron 1965"
: Washington National Records Center, Suitland, Md.
Record Group 255
Accession Number 79-0649
Box 25
Folder Number 35
"Title of file folder as it appears on original"

NAS:
: National Academy of Sciences Archives, 2101 Constitution Avenue, Washington, D.C. The rest of the citation following the colon gives the identification of the file folder as it appears in the archives.

JPLHF
: Jet Propulsion Laboratory History Files, JPL Library, Pasadena, Calif. The numbers following designate a folder in the history files.

AR
: Active Records, still in the possession of the person cited when they were consulted.

John Smith, OHI by Joe Jones 1 Jan. 1970 (SAOHP), p. 99.
: Oral History Interview of John Smith by Joe Jones on January 1, 1970, page 99 of transcript located in the Space Astronomy Oral History Project, National Air and Space Museum, Washington, D.C.

SAOHPWF
: Space Astronomy Oral History Project Working Files. These files contain curriculum vitae, publications, and in some cases unpublished materials donated by the interview subject.

APPENDIX 2: PLANETARY EXPLORATION–
FLIGHT PROGRAM SUMMARY

Planet	Spacecraft	Launch Date	Encounter
Venus	Mariner 2	8/27/62	F: 12/14/62
Mars	Mariner 4	11/28/64	F: 7/14/65
Venus	Mariner 5	6/14/67	F: 10/19/67
Mars	Mariner 6	2/25/69	F: 7/31/69
Mars	Mariner 7	3/27/69	F: 8/5/69
Mars	Mariner 9	5/30/71	O: 11/13/71–10/27/72
Jupiter	Pioneer 10	3/3/72	F: 12/4/73
Jupiter	Pioneer 11	4/6/73	F: 12/5/74
Saturn	Pioneer 11		F: 9/1/79
Venus	Mariner 10	11/3/73	F: 2/5/74
Mercury	Mariner 10		F: 3/29/74
			F: 9/21/74
			F: 3/16/75
Mars	Viking 1	8/20/75	O: 6/21/76–8/7/80
			L: 7/20/76– -/-/82
Mars	Viking 2	9/9/75	O: 8/7/76–7/25/78
			L: 9/3/76–4/12/80
Venus	Pioneer-Venus 1	5/20/78	O: 12/4/78
	Pioneer-Venus 2	8/8/78	P: 11/-/78
Jupiter	Voyager 1	9/5/77	F: 3/5/79
Saturn	Voyager 1		F: 11/12/80
Jupiter	Voyager 2	8/20/77	F: 7/9/79
Saturn	Voyager 2		F: 8/25/81
Uranus	Voyager 2		F: 1/21/86
Neptune	Voyager 2		F: 8/-/89

F = Flyby encounter
O = Orbiter, from date it entered orbit to date it ceased sending data
L = Lander, from touchdown to shutdown
P = Probe sent into atmosphere
Mariner 10 remained in solar orbit and made multiple encounters of Mercury

APPENDIX 3: NASA POLICY ON SUPPORT TO GROUND-BASED ASTRONOMY

Mar 23 1965

THE ADMINISTRATOR

Dr. Leland J. Haworth
Director
National Science Foundation
Washington, D.C.

Dear Dr. Haworth:

The purpose of this letter is to transmit the NASA assessment of the Whitford Report and plans for the support of Ground-Based Astronomy to the National Science Foundation (NSF) for use by Dr. Keller and his group in developing a government-wide assessment of the report.

Enclosed is a review of the report prepared by Dr. Nancy Roman based on the advice and recommendation of the NASA Astronomy Subcommittee and Drs. Liddel, Brunk, Holloway, Smith, and Mr. Scott—those persons in NASA with the primary responsibility for developing the Astronomy Program and providing the supporting facilities.

In general we assess the Whitford Report as an accurate and thorough analysis, from the viewpoint of astronomers, of the rationale and needs of ground-based astronomy for the next decade. We agree with the Committee on Science and Technology that the recommended program is conservative and may not provide sufficient opportunities for young astronomers. In addition, we think that they may have underestimated the interest in astronomy which will be generated by the results from the space program. We will not be able to accurately assess the impact of space astronomy until after one or more Orbiting Astronomical Observatories (OAO's) have flown.

We endorse the Whitford Report as a statement of the needs of stellar and galactic astronomy, both optical and radio. However, it explicitly omits detailed consideration of the requirements of solar astronomy. A study of the sun and its influence on the earth and interplanetary space is a cornerstone of our country's exploration of the solar system and of the earth as a planet. Dr. Henry Smith has prepared an estimate of the future ground-based facilities required for solar astronomy for use by Dr. Keller's committee. That report reflects our view that additional large, intermediate and small optical solar telescopes are required for ground-based solar astronomy to support this important area of science.

We recognize the close connection between ground-based astronomy and the exploration of space; however, we feel that the broad general support of ground-based astronomy should continue to be the responsibility of the NSF. NASA has in the past supported certain ground-based facilities and will continue to do so in the future after coordination with other government agencies to avoid duplication. The NASA funds available for the support of ground-based astronomy can support only a fraction of those new ground-based facilities which do one or more of the following:

1. Contribute directly to understanding the results from the space program.

2. Assist in the conception, design, development and testing of instruments and spacecraft.

3. Perform the exploratory and basic research necessary to guide the planning and evaluation of scientific missions and experiments in space.

In addition, ground-based facilities at academic institutions will increase the number of graduate astronomers which are available to participate in the space program.

Enclosed is a description of the major ground-based astronomical facilities which NASA is supporting. This list is broken into those facilities which have been approved, those which we are likely to approve as soon as we have completed current discussions with the institutions involved, and an estimate of the facilities which we plan to support in the future. Our plans are of course dependent on future Congressional action and upon the questions which arise and the discoveries which are made in the flight program—"the opportunities which beckon."

The facilities which are in Enclosure 2 satisfy one or more of the above criteria; have been coordinated with other government agencies to avoid duplication; and are usually located at academic institutions which train astronomers.

In addition to facilities we support astronomical education through:

1. Encouragement of the award of NASA training grants to doctoral astronomy students.

2. Continued use of pre- and post-graduate student assistants on research projects of interest to NASA.

3. Support and encouragement of conferences and symposia for professionals and summer courses for undergraduate and graduate students conducted by NASA centers in astronomy and related areas.

As indicated in Dr. Roman's report, we are making the result of our research on large radio dishes and arrays of such dishes freely available to the astronomical community. We have also reviewed the possibility of making the NASA Tracking and Data Acquisition large dish facilities available for radio astronomy observations on a non-interference basis with support of the space flight program. However, the workload on our existing facilities and the anticipated workload of the 210-foot dish currently under construction at Goldstone, California, are such that only a small amount of radio astronomy work could be undertaken. That which could be undertaken would not have a significant impact on the total radio astronomy effort envisioned in the Whitford Report.

Sincerely yours,

Hugh L. Dryden
Deputy Administrator

Enclosures
1. Memo Roman to Newell dtd 3/16/65
2. Table—"Facilities Which Have Been Approved"
3. Table—"Planned Facilities"
4. Table—"Estimated Future Level of Support of Ground-Based Astronomy Facilities"

Location	Aperture (in.)	Type*	Owner/Operator Principal Investigator	Completion Date° Grant Number	Remarks
Mt. Palomar, Calif.	200	P-C-D	Caltech R. B. Leighton	(1948) NsG-426	Data-handling system and focal instrumentation only
Mt. Locke, Tex.	107	P-C-D	Texas H. J. Smith	1963–1969 NASr-242	Includes spectrograph and Connes-type interferometer
Mt. Wilson, Calif.	100	N-C-D	Carnegie R. B. Leighton	(1917) NsG-426	Data-handling system and focal instrumentation only
Mauna Kea, Hawaii	88	C-D	Hawaii J. T. Jeffries	1965–1970 NSr-2-1-9	Includes Connes-type interferometer
Mt. Locke, Tex.	82	P-C-D	Texas H. J. Smith	(1939)	Resurfaced mirror and improved coudé spectrograph
Catalina, Ariz.	61	C	Arizona LPL G. P. Kuiper	1962–1965 NASr82	First major planetary telescope funded by NASA
Mt. Wilson, Calif.	60	N-C-D	Carnegie H. W. Babcock	(1908) 1965–1970 NSr9-140-001	Modernized for rapid changeover from stellar to planetary observations
Catalina, Ariz.	60	C	NASA	1969	Huntsville, Goddard Institute for Space Studies, SUNY, Arizona
Mt. Palomar, Calif.	60	C-D	Carnegie	1971	Mirror only funded by NASA
Bethany Sta., Conn.	40	P-C-D	Yale D. Brouwer	1966 NsG-29	Astrometric
Palestine, Tex.	36	C	Princeton M. Schwarzschild	1963–1970	Stratoscope II flights, NSF/ONR/NASA
Greenbelt, Md.	36	C-D	NASA	1967	Goddard Space Flight Center test telescope
Flagstaff, Ariz.	30	P-C	USGS E. M. Shoemaker	1964 R-66	Fund transfer, E. Shoemaker, lunar mapping

Location	Size	Type	Institution	Year/Grant	Description
Palestine, Tex.	28	C	Arizona LPL, T. Gehrels	1966 NASr-138	Polariscope balloon-borne telescope flights
Table Mtn., Calif.	24	C-D	JPL	1966	Coudé and spectrograph, incl. Connes-type interferometer
Ojai, Calif.	24	P-C	UCLA, L. Aller	1965 NsG-237	Bent Cassegrain on multidisciplinary grant to Willard Libby
White Mtn., Calif.	24	P-C	Caltech, H. Brown	NsG-56	Identical to above, on multidisciplinary grant
Las Cruces, N.Mex.	24	C	Northwestern, J. A. Hynek	1965 NsG-597	Image orthicon for LTP patrol
Las Cruces, N.Mex.	24	C	New Mexico State, C. Tombaugh	1967 NsG-142	Photographic Planetary Patrol
Mauna Kea, Hawaii	24	C	NASA-Lowell	1969	International Planetary Patrol (IPP) cameras & filters only
Cerro Tololo, Chile	24	C	NASA-Lowell	1969	IPP
Australia	24	C	NASA-Lowell	1969	IPP cameras & filters only
India	24	C	NASA-Lowell	1969	IPP cameras & filters only
South Africa	24	C	NASA-Lowell	1969	IPP cameras & filters only
Las Cruces, N.Mex.	16	C	New Mexico State, C. Tombaugh	1962 NsG-142	Modified used instrument
Table Mtn., Calif.	16	C	JPL	1963	Modified used Nishaimura

Source: NASA Planetary Astronomy Office Files
Table adapted from *Astronomy and Astrophysic for the 1970s.*
(Washington, D.C.: NAS, 1972), volume 2, pp. 388–393.

*P=Prime Focus; C=Cassegrain; D=Coudé; N=Newtonian Reflector

°In cases of modification only, dates in parentheses refer to completion of original instrument. Ranges of dates refer to start of grant and completion of instrument.

Title to telescope held by receiving institution; contract includes operational support, clauses guaranteeing use of instrument for lunar and planetary research for a specified percentage of time, and access by PIs on other NASA lunar and planetary grants and contracts. Specific arrangements vary.

APPENDIX 5: NASA RESEARCH GRANTS AND CONTRACTS FOR PLANETARY ASTRONOMY, 1959–1963

Start Date	Grant Number	University	Principal Investigator	Project Description
5/8/59	NsG8	Yale	Lilly & Brouwer	Solar parallax using absorption line spectrum
5/15/59	NsG4	Michigan	Siegel	Radar planetary mapping
7/59	HS149	Yale	Brouwer	Summer Institute in Dynamical Astronomy
8/1/59	NsG29	Yale	Brouwer	Celestial mechanics
8/1/59	NsG37	Chicago	Kuiper	Lunar astronomy
10/1/59	NsG43	Cincinnati	Herget	Celestial mechanics
12/1/59	NsG56	Caltech	H. Brown	Multidis. grant incl. lunar and planetary research
12/1/59	NsG58	Maryland	Singer/Opik	Theoretical investigations
12/1/59	NsG69	Princeton	Schwarzschild	Stratoscope balloon observatory
1/1/60	NsG64	Harvard	Menzel	IR lunar radiometry
3/1/60	NsG73	Yale	H. Smith	Jupiter radio astronomy
6/15/60	NsG89	Harvard	Menzel	Planetary multicolor photometry/polarimetry
1/1/61	NsG118	Chicago	Chamberlain	Theoretical study of planetary atmospheres
2/1/61	NsG129	U. New Mex.	Moore	Planetary radar
4/15/61	NsG142	N. Mex. St.	Tombaugh	Photographic planetary patrol
6/1/61	NsG161	Arizona	Kuiper	Lunar and planetary studies—multiple
8/1/61	NsG181	Michigan	Haddock	Space probe radio experiments
8/12/61	NGL3-2-2	Arizona	J. C. Smith	Planetary astronomy
10/1/61	NsG234	MIT	Reintjes	Venus radar
11/1/61	NsG213	Ohio St.	Lewis	Radar astronomy

Date	Grant	Institution	Investigator	Topic
1/1/62	NsG223	Arizona	Meinel	Planetary spectroscopy
2/1/62	NsG224	Fla. St.	Barrow	Jupiter radio astronomy
4/1/62	NsG255	Berkeley	Sagan	Stratoscope IR Mars observations
5/31/62	NASR113	Princeton	Schwarzschild	Modify Stratoscope II for IR Mars Observations
8/1/62	NsG280	Illinois	J. H. Barlett	Celestial mechanics
—	NsG291	Harvard	Whipple	Photographic meteor patrol
8/15/62	NsG297	Manchester	Kopal	Lunar observations from Japan
8/22/62	NASR82	Arizona	Kuiper	60-inch reflector
11/1/62	NASR138	Arizona	Gehrels	Develop spacecraft polarimeter (later balloon-borne)
3/20/63	NsG362	Georgetown	Heyden	Observational and laboratory spectroscopy
4/1/63	NsG432	Texas	Tolbert	Radio astronomy
4/29/63	NGL44-12-6	Texas	W. J. Wilson	Radio astronomy
5/8/63	NsG407	Yale	Douglas	Jupiter radio astronomy
—	NsG419	MIT	Barrett	Microwave solar system astronomy
—	NsG426	Caltech	Leighton	Multidis. grant incl. planetary research
5/20/63	NGL5-2-7	Caltech	Neugebauer	Infrared spectroscopy
7/1/63	NsG451	Lowell	Hall	Planetary astronomy
7/3/63	NsG493	Arizona	Gehrels	Planetary photometry
10/28/63	NsG(F)15	Lowell	Hall	Planetary Research Center construction

Source: NASA Semiannual Reports to Congress
NASA University Affairs Office Reports
NASA Planetary Astronomy Office Files

Listing does not include Principal Investigators contracted by NASA Centers and JPL, research performed in-house, contracts with private firms, or fund transfers to other agencies for related research.
Principal Investigator as listed on grant or contract often represents several other investigators not listed.

151

Notes

INTRODUCTION

1. The remark was quoted by Gerard De Vaucouleurs, *Proceedings of the Lunar and Planetary Exploration Colloquium* II/1 (23–24 Sept. 1959): 53; Hibbs's original remark was in March. Ibid., I/5 (18 Mar. 1959): 28.

2. Ibid., II/1: 53. Jaffe's and De Vaucouleurs's remarks were made in reply to a suggestion that it was simply a question of providing financial incentive.

3. Ibid.

4. Walter McDougall, *The Heavens and the Earth: A Political History of the Space Age* (New York: Basic Books, 1985).

5. Mark Washburn, *Distant Encounters: The Exploration of Jupiter and Saturn* (New York: Harcourt Brace Jovanovich, 1983), p. 221.

6. Quoted in Washburn, *Distant Encounters,* p. 222.

ONE. PLANETARY ASTRONOMY IN THE AGE OF ASTROPHYSICS (1900–1958)

1. Loren Graham, Wolf Lepenies, and Peter Weingart, eds., *Functions and Uses of Disciplinary Histories* (Dordrecht: Reidel, 1983).

2. Gerard P. Kuiper, "Preface," *Planets and Satellites* (Chicago: University of Chicago Press, 1961), Volume III of G. P. Kuiper and B. M. Middlehurst, eds., *The Solar System* (in five volumes). Only four volumes of this series, intended to be a comprehensive review of the state of solar system astronomy, were eventually published.

3. Walter Fricke, "New Impetus to the Exploration of the Solar System," inaugural address in W. Fricke and G. Teleki, eds., *Sun and Planetary System* (Dordrecht: Reidel, 1982), p. 9.

4. Neville J. Woolf, "The Impact of Space Studies on Astronomy," in William E. Frye, ed., *Impact of Space Exploration on Society* (Tarzana, Calif.: American Astronautical Society, 1966), pp. 180–81.

5. Historian Stephen Brush has claimed that planetary science's prestige was eclipsed during the first half of the century, so much so that "many competent scientists who might well have made important discoveries in planetary science did not even consider entering this field." Stephen G. Brush, "Planetary Science: From Underground to Underdog," *Scientia* 113 (1978): 771–87. See also his "Nickel for Your Thoughts: Harold Urey and the Origin of the Moon," *Science* 217/4563 (3 Sept. 1982): 891–98, and "Harold Urey and the Moon: The Interaction of Science and the Apollo Program," pp. 437–70 in Peter M. Bainum and Dietrich E. Koelle, eds., *Spacelab, Space Platforms, and the Future: Proceedings of the Twentieth Goddard Memorial Symposium* (San Diego, Calif.: American Astronautical Society, 1982); George E. Webb, *Tree Rings and Telescopes: The Scientific Career of A. E. Douglass* (Tucson: University of Arizona Press, 1983).

6. Agnes M. Clerke, *Popular History of Astronomy During the Nineteenth Century,* 2nd ed. (Edinburgh: Adam and Charles Black, 1887), p. 2.

7. The development of stellar and galactic astronomy during this period is treated in a number of general histories; see, among others: A. O. Pannekoek, *A History of Astronomy* (London: Allen & Unwin, 1961), and Otto Struve and Velta Zebergs, *Astronomy of the Twentieth Century* (New York: Macmillan, 1962). For additional citations see David H. DeVorkin, *The History of Modern Astronomy and Astrophysics: A Selected, Annotated Bibliography* (New York: Garland, 1982). More specialized historical studies on the development of stellar astronomy during

this period may be found in Michael Hoskin, *Stellar Astronomy: Historical Studies.* (New York: Science History Publications, 1983); Karl Hufbauer, "Astronomers Take Up the Stellar Energy Problem, 1917–1920," *Historical Studies in the Physical Sciences* 11/2 (1981); David H. DeVorkin, "A Sense of Community in Astrophysics: Adopting a System of Stellar Spectral Classification," *Isis* 72 (1981): 29–49; DeVorkin, "The Origins of the Hertzsprung-Russell Diagram," pp. 61–78 in A. G. D. Philip and David H. DeVorkin, eds., *In Memory of Henry Norris Russell* (Albany, N.Y.: 1977); DeVorkin and Ralph Kenat, "Quantum Physics and the Stars: (I) The Establishment of a Stellar Temperature Scale," *Journal for the History of Astronomy* 14 (1983): 102–32.

Galactic astronomy and the new cosmology of the expanding universe is treated in, among others: Robert W. Smith, *The Expanding Universe: Astronomy's Great Debate, 1900–1931* (Cambridge: Cambridge University Press, 1982), and Richard Berendzen, *Man Discovers the Galaxies* (New York: Science History Publications, 1976).

8. Rudolph Thiel, *And Then There Was Light* (New York, 1957), p. 243, quoted in William Graves Hoyt, *Lowell and Mars* (Tucson: University of Arizona Press, 1976), p. 8.

9. The story of Percival Lowell and his observatory is treated extensively in Hoyt, *Lowell and Mars* and *Planets X and Pluto* (Tucson: University of Arizona Press, 1980).

For discussion of Lowell's ambiguous status in the astronomical community see Norriss S. Hetherington "Percival Lowell: Professional Scientist or Interloper?" *Journal of the History of Ideas* 42 (1981): 159–61, and "Amateur versus Professional: The British Astronomical Association and the Controversy over Canals on Mars," *Journal of the British Astronomical Association* 86/4 (1976): 302–308. Although Lowell was a central figure, controversy over the proper role of professional and amateur, appropriate type and use of telescopic instrumentation, and the validity of various planetary observations was confined neither to Lowell nor to Mars; see John Lankford, "Amateur versus Professional: The Transatlantic Debate over the Measurement of Jovian Longitude," *Journal of the British Astronomical Association* 89 (1979): 574–82, and "Amateurs versus Professionals: The Controversy over Telescope Size in Late Victorian Science," *Isis* 72 (Mar. 1981): 11–28. On the professionalization of astronomy at the turn of the century involving astrophysics and the role of instrumentation see John Lankford, "Amateurs and Astrophysics: A Neglected Aspect in the Development of a Scientific Specialty," *Social Studies of Science* 11 (1981): 275–303, and Marc Rothenberg, "Organization and Control: Professionals and Amateurs in American Astronomy, 1899–1918," *Social Studies of Science* 11 (1981): 305–25.

10. *A Hundred Years of Astronomy* (New York: Macmillan, 1938), p. 147, quoted in Struve and Zebergs, p. 147.

11. Hoyt, *Lowell and Mars*, pp. 43–50, 89–94, 224–25, 237–39, 293–97.

12. Hoyt, *Lowell and Mars*, pp. 173–86, 190–200.

13. Hoyt, *Lowell and Mars*, pp. 302–305.

14. Ibid.

15. Struve and Zebergs, pp. 141, 147.

16. I. S. Shklovskii and Carl Sagan, *Intelligent Life in the Universe* (San Francisco: Holden-Day, 1966), p. 276. I thank Michael J. Crowe for bringing these and other citations to my attention. See his *The Extraterrestrial Life Debate, 1750–1900: The Idea of a Plurality of Worlds from Kant to Lowell* (New York: Cambridge University Press, 1986).

17. Clyde W. Tombaugh and Patrick Moore, *Out of the Darkness: The Planet Pluto* (Harrisburg, Pa.: Stackpole Books, 1980), p. 101.

18. Hoyt, *Planets X*, pp. 144–45.

19. Tombaugh, *Out of the Darkness*, p. 101.

20. Hoyt, *Planets X*, pp. 155, 147.

21. Otto Struve and Velta Zebergs, p. 17.

22. David H. DeVorkin, "The Maintenance of a Scientific Institution: Otto Struve, the Yerkes Observatory, and Its Optical Bureau During the Second World War," *Minerva* 18/4 (Winter 1980): 595–623; see especially p. 595.

23. Bernard Lovell, "The Effect of Defence Science on the Advance of Astronomy," *Journal for the History of Astronomy* VIII (1977): 151, 167.

24. Daniel J. Kevles, *The Physicists: The History of a Scientific Community in Modern America* (New York: Random House, 1979), chapter 21, esp. pp. 341–42; Harvey M. Sapolsky, "Science, Technology, and Military Policy," pp. 443–71 in Ina Spiegel-Rosing and Derek de Solla Price, eds., *Science, Technology, and Society: A Cross-Disciplinary Perspective* (Beverly Hills, Calif.: Sage, 1977), pp. 446–47, 458–59; Sapolsky, "Academic Science and the Military: The Years Since the Second World War," pp. 379–99 in Nathan Reingold, ed., *The Sciences in the American Context: New Perspectives* (Washington, D.C.: Smithsonian Institution Press, 1979), pp. 383–85.

25. Earl C. Slipher, Project Director, *The Study of Planetary Atmospheres: Final Report*, U.S. Air Force Contract AF 19(122)-162, 30 Sept. 1952, p. 2.

26. On the history and fortunes of the Lowell Observatory during the 1950s see Ronald E. Döel, *Unpacking a Myth: Interdisciplinary Research and the Growth of Solar System Astronomy in America, 1920–1958* (Princeton University Ph.D. diss., 1989).

27. An increased interest in meteorology also developed in the aftermath of the First World War. "New means of warfare, such as long-range artillery bombardments, gas attacks, and especially aerial combat, had already led to a mobilization and rapid expansion of meteorological services. Postwar expectations for rapid development of air transport similarly prompted meteorologists to promote the growth of their science, since new predictive methods and greater understanding of atmospheric phenomena would be required to insure the safe operations of aerial routes." Robert Marc Friedman, "Constituting the Polar Front, 1919–1920," *Isis* 73/268 (Sept. 1982): 343–62, esp. p. 348.

28. David Bushnell, *The Sacramento Peak Observatory, 1947–1962.* (Washington, D.C.: U.S. Air Force Office of Aerospace Research, Historical Division, 1962). Bushnell traces the origins of the project to the Climax, Colorado station of the Harvard College Observatory established by Donald Menzel and Walter Orr Roberts in 1940. Menzel and Roberts did the site survey, and in 1948 a regular program of solar observations began at Sacramento Peak, with portions of the work contracted to both Climax and Harvard.

On space technology and solar science see Karl Hufbauer, *Exploring the Sun* (forthcoming).

29. Clayton S. White and Otis O. Benson, Jr., eds., *Physics and Medicine of the Upper Atmosphere* (Albuquerque: University of New Mexico Press, 1952). This symposium gathered together astronomers, geophysicists, engineers, medical doctors, and other specialists to consider the problems which needed to be solved since recent and anticipated developments in aviation "have currently projected our flying personnel into the immediate vicinity of an environmental frontier beyond which our knowledge is incomplete, and, in some respects, totally lacking" (p. xiv), a frontier "opened by the advent of rocket and atomic power" (p. xvii).

An excellent review of some of the early proposals is R. Cargill Hall, "Early U.S. Satellite Proposals," *Technology and Culture* 4 (Fall 1963): 410–34; also valuable is Constance McLaughlin Green and Milton Lomask, *Vanguard: A History* (Washington, D.C.: NASA SP-4202, 1970 and Smithsonian Institution Press), chapter 1.

30. For a discussion of various types of missile guidance developed by the Jet Propulsion Laboratory for the Army in the 1950s see Clayton R. Koppes,

JPL and the American Space Program: A History of the Jet Propulsion Laboratory (New Haven: Yale University Press, 1982). The development of reliable all-inertial guidance systems in the late 1950s allowed the trajectory to be controlled on board but of course did not lessen the importance of understanding the various effects of the atmosphere on radio communications and trajectories.

31. Nick A. Komons, *Science and the Air Force: A History of the Air Force Office of Scientific Research* (Arlington, Va.: Office of Aerospace Research, 1966), p. 113. Scientific advisory bodies and public reaction had ended any but intellectual consideration and planning for satellites and space exploration in the late 1940s. The "manned bomber" contingent of the Air Force had gained the upper hand, but with the development of powerful but lightweight nuclear warheads, and the awareness of parallel developments in the Soviet Union in the early 1950s, the Air Force and Army had accelerated development of ballistic missiles.

32. Morton Alperin and Marvin Stern, eds., *Vistas in Astronautics: First Annual Office of Scientific Research Astronautics Symposium* (New York: Pergamon, 1958).

33. Morton Alperin and Hollingsworth F. Gregory, eds., *Vistas in Astronautics, Volume II: Second Annual Astronautics Symposium* (New York: Pergamon, 1959). Similar lunar and planetary interests can be found in Charles T. Morrow, et al., eds., *Proceedings of the Fourth Air Force Ballistic Missile Division/Space Technology Laboratories Symposium: Advances in Ballistic Missile and Space Technology* (New York: Pergamon, 1960), 3 vols.

34. *Aviation Week* 68/13 (31 Mar. 1958): 22–23; JPL's Al Hibbs later recalled: "We saw no military use of the moon or planets. And we were deeply concerned that some Air Force officer might try to put that one over, and we wanted to make sure that that was shot down very early." Hibbs, OHI by Needell and Tatarewicz, 9 Dec. 1982, p. 34, and letter to the author, 2 May 1983, pp. 1–2.

35. *Aviation Week* 68/26 (30 June 1958): 20.

36. *Aviation Week* 69/13 (29 Sept. 1958): 18.

37. *Aviation Week* 70/17 (27 Apr. 1959): 27–28.

38. *Aviation Week* 7/13 (28 Sept. 1959): 26–27.

39. James R. Killian, *Sputnik, Scientists, and Eisenhower: A Memoir of the First Special Assistant to the President for Science and Technology* (Cambridge, Mass.: MIT Press, 1977), p. 128; on the interservice rivalries for space see pp. 122–23.

40. Komans, *Science and the Air Force*, p. 151.

41. Brig. Gen. Benjamin G. Holzman later described the area of the Air Force's interest as "the space environment," which he defined as beginning in the center of the earth (where the magnetic field originates) and extending to the "farthest-known part of the universe." Holzman's explanations of the propriety of the Air Force being concerned with so wide a realm are enlightening. See Holzman, "The Space Environment," in *The U.S. Air Force in Space* (New York: Praeger, 1966), esp. pp. 54–55.

42. The need for mission justification for civilian contracted research, "painting projects blue" as it was called in ONR, did have its ups and downs. See Sapolsky, "Academic Science and the Military," pp. 388–90.

43. Interservice rivalries had prevented Navy Bureau of Aeronautics and Army Air Force collaboration on an earth satellite project as early as 1945 because "General Curtis Lemay resented the Navy invasion into a field which was the province of the Army Air Force," Green and Lomask, *Vanguard*, pp. 7–9. A decade later, but still before Sputnik, similar problems prevented Army Ballistic Missile Agency and Navy collaboration on the International Geophysical Year earth satellite program, since "breaking the space barrier would be an easier task than breaking the interservice barrier." Clifford C. Furnas, quoted in *Vanguard*, p. 51.

44. Charles L. Stewart, quoted in Wernher Von Braun and Frederick I. Ordway, III, *History of Rocketry and Space Travel* (New York: Crowell Co., 1966), p.

116. William Sims Bainbridge, *The Spaceflight Revolution: A Sociological Study* (New York: Wiley, 1976), argues that contrary to the popular wisdom that governments manipulated the rocketry pioneers such as Von Braun to their own ends, it was actually the rocketry pioneers who manipulated the military and government into providing the resources to realize their dreams. But see Walter A. McDougall's reservations, "Technocracy and Statecraft . . . ," pp. 1013–14.

45. Wernher Von Braun, *The Mars Project* (Urbana: University of Illinois Press, 1953), pp. 1, 3. Translation of *Das Marsprojekt,* special issue of the magazine *Weltraumfahrt* (Esslingen, Germany: Bechtle Verlag, 1952). NASA advanced planning studies in the late 1950s for Mars expeditions bore a striking resemblance to the plans contained in this study; see Edward C. Ezell, "Man on Mars: The Mission That NASA Did Not Fly," *STTH: Science, Technology, and the Humanities* 3 (1980): 21–34.

46. John M. Logsdon, *The Decision to Go to the Moon: Project Apollo and the National Interest* (Cambridge, Mass.: MIT Press, 1970), pp. 49–51; Edward C. Ezell and Linda N. Ezell, *On Mars: Exploration of the Red Planet, 1958–1978* (Washington, D.C.: NASA SP-4212, 1984), pp. 4–10; Comment Draft, Janaury 1979, chapter 1, pp. 5–11.

47. Logsdon, *Decision,* pp. 49–51; Ezell and Ezell, *On Mars,* chapter 1, pp. 5–11.

48. Clyde W. Tombaugh, *Interim Report on Search for Small Earth Satellites for the period 1953–56* (Physical Science Laboratory, State College, New Mexico), sponsored by Office of Ordnance Research, U.S. Army Project 5B99-01-004, pp. 4, 21–22, and *Search for Small Earth Satellites. Final Technical Report. 30 June 1959.*

49. Tombaugh's New Mexico State project and Brad Smith's later career in both ground-based and space-based planetary research will be discussed in a later chapter. Charles F. Capen, one of the ablest visual planetary observers, later worked on the survey for the Air Force's proposed Cloudcroft facility, was the first full-time observer at Jet Propulsion Laboratory's Table Mountain (planetary) observatory, and later returned to Lowell Observatory to work on the International Planetary Patrol conducted by NASA in support of Mariner and Viking probes.

50. Harvey M. Sapolsky, "Academic Science and the Military: The Years Since the Second World War," pp. 379–99, in Nathan Reingold, ed., *The Sciences in the American Context: New Perspectives* (Washington, D.C.: Smithsonian Institution Press, 1979), pp. 383–85. For civilian wartime science organization see Carroll Pursell, "Science Agencies in World War II: The OSRD and Its Challengers," pp. 359–78 in the same volume.

51. *ONR Research Reviews* (1950).

52. Sapolsky, "Academic Science and the Military," p. 382. For the development of radar at NRL see David K. Allison, *New Eye for the Navy: The Origin of Radar at the Naval Research Laboratory* (Washington, D.C.: NRL/GPO, 1981).

53. Cornell H. Mayer, "The Temperatures of the Planets," *Scientific American* 204/58 (May 1961): 58–65; Alex G. Smith and Thomas D. Carr, *Radio Exploration of the Planetary System* (New York: Van Nostrand, 1964), pp. 55–60. Speaking of the discovery of decimetric radiation from Jupiter, an anonymous astronomer (quoted in David Edge and Michael Mulkay, *Astronomy Transformed: The Emergence of Radio Astronomy in Britain* [New York: Wiley, 1978], p. 280) said: "It's difficult to say to what extent it was radio astronomy and to what extent it was satellites that really renewed interest in the planets. . . . With optical telescopes on the ground they had reached the end of their resources. There had been a tremendous interest in it in the early part of this century and then the techniques wouldn't take them any further; and then radio astronomy and then satellites enabled this field to start again."

54. Malcolm D. Ross, "Balloon Astronomy," *ONR Research Reviews* (May 1958): 5–12, and "Plastic Balloons for Planetary Research," *Journal of the Astronautical Sciences* 5 (Spring 1958): 5–10. The unmanned version, Stratoscope I, under the direction of Princeton astronomers Martin Schwarzschild and Robert E. Danielson, achieved high-resolution photographs of the sun from 1957 to 1959. NSF and NASA joined in funding an improved *Stratoscope II* which took infrared spectra of Mars in 1963 and, after some development difficulties, high-resolution imagery of Jupiter and Uranus in 1970.

The manned version, Strato-Lab, was to be used by Johns Hopkins University astronomer John Strong (with Malcolm Ross as pilot) to take infrared spectra of Mars in the fall of 1958, but the balloon ruptured. The equipment was later used in November 1959 in a successful attempt to detect water vapor in the atmosphere of Venus. See John Strong, "Observations with Satellite-Substitute Vehicles," pp. 85–89 in William W. Kellogg and Carl Sagan, eds., *The Atmospheres of Mars and Venus: A Report by the Ad Hoc Panel on Planetary Atmospheres of the Space Science Board* (Washington, D.C.: NAS, 1961); Strong, "Infrared Astronomy by Balloon," *Scientific American* 212 (Jan. 1965): 28–37. Strong's system was improved and later used in an Air Force Cambridge Research Laboratory unmanned balloon program, BALAST. AFCRL also sponsored Northwestern University astronomer J. Allen Hynek's manned project STAR GAZER, as well as an in-house effort directed by John Salisbury, project SKY TOP, aimed at infrared radiometry of the moon, Mars, and Venus. See files of press releases, clippings, and articles in SAOHPWF; *Proceedings of the AFCRL Balloon Symposium* (AFCRL Report 63–919, Dec. 1963); and *Proceedings of the AFCRL Scientific Balloon Symposium* (Annual, 1964–75).

55. Ezell and Ezell, *On Mars*, pp. 31–32.

56. John Naugle, "Planning Planetary Missions," lecture delivered at the Seminar on Space Policy, National Air and Space Museum, 4 Mar. 1982 transcript in SAOHPWF; Oran W. Nicks, OHI by Tatarewicz 2 Aug. 1983.

A bewildering assortment of launch vehicles of various sizes and capabilities was already in existence or in development before Sputnik. The chief missiles potentially capable of delivering payloads beyond earth orbit were the Air Force's Atlas (combined with high energy upper stages such as Centaur), the Army's Juno, Thor, and Saturn (based on Von Braun's V-2), and an emerging concept at Jet Propulsion Laboratory for a high-energy vehicle with interplanetary capability called Vega. See Richard P. Hallion, "The Development of American Launch Vehicles since 1945," pp. 115–34 in Paul A. Hanle and Von Del Chamberlain, eds., *Space Science Comes of Age: Perspectives in the History of the Space Sciences* (Washington, D.C.: Smithsonian, 1981), and chapter 10 of Homer E. Newell, *Beyond the Atmosphere: Early Years of Space Science* (Washington, D.C.: NASA SP-4211, 1981). For the development problems of JPL's Vega see Clayton R. Koppes, *JPL and the American Space Program: A History of the Jet Propulsion Laboratory* (New Haven: Yale University Press, 1982).

57. Contracting for advanced spacecraft systems and development of scientific instruments required a significant corporate scientific capability. The development of such divisions, and their problematic status between "pure" and "applied" science, is a fertile field for study.

58. *Proceedings of the Lunar and Planetary Exploration Colloquium*, Colloquiums 1–13, May 1958–May 1963 (published and distributed by Missile Division, North American Aviation, Inc.). See the individual numbers for representative attendance lists. The scientific discussions at the colloquia relating to ground-based planetary astronomy will be discussed below.

59. Charles E. Anderson, *Physical Properties of the Planet Venus* (Santa Monica, Calif.: Douglas Aircraft Missile and Space Systems Division, Report SM-41506,

July 1962); Anderson, *Physical Properties of the Planet Mars* (Santa Monica, Calif.: Douglas Aircraft Missile and Space Systems Division, Report SM-43634, Aug. 1963).

60. Walter A. McDougall, "Technocracy and Statecraft in the Space Age: Toward the History of a Saltation," *American Historical Review* 87/4 (Oct. 1982): 1035.

61. Nancy Roman to Roger C. Moore, 23 Oct. 1961, WNRC 255-74-663 (48): "Dec. 1959–1961, Astronomy and Solar Physics."

62. Letter, S. Fixler to E. M. Cortright, Director, Lunar and Planetary Programs, 9 May 1961 and reply to Fixler from F. D. Kochendorfer, 27 June 1961, WNRC 67A601(7): "Astronomy."

63. G. P. Kuiper, "Discourse following Award of Kepler Gold Medal at A.A.A.S. Meeting, Dec. 28, 1971," *Communications of the Lunar and Planetary Laboratory,* no. 183. The book was apparently James Jeans, *The Mysterious Universe* (Cambridge: Cambridge University Press, 1930), reviewed by Kuiper in *Hemel en Dampkring* 29 (1930): 33–36 and 30 (1932): 69.

64. Northwestern University physicist Samuel A. Goudsmit describes the mission from his role as scientific director in *ALSOS* (New York: Schuman, 1947). The overall mission was directed by Gen. Leslie Groves from the Intelligence Division of the War Department.

65. Kuiper, "Discourse," op. cit.; "German Astronomy during the War," *Popular Astronomy* LIV/6 (June 1946): 263–86.

66. Dale P. Cruikshank, "20th-Century Astronomer," *Sky and Telescope* 47 (Mar. 1974): 159–64, and "Gerard Peter Kuiper," *Biographical Memoirs of the National Academy of Sciences* (in press).

67. G. Herzberg, "Laboratory Absorption Spectra Obtained with Long Paths," chapter 13 in G. P. Kuiper, ed., *The Atmospheres of the Earth and Planets* (Chicago: University of Chicago Press, 1952), pp. 406–16.

68. G. P. Kuiper, "Carbon Dioxide on Mars," *Harvard Announcement Card* 851 (1947); "New Absorptions in the Uranus Atmosphere," *Astrophysical Journal* 109 (1949): 540–41; "The Diameter of Neptune," *Astrophysical Journal* 110 (1949): 93–94; "Object Near Neptune," *IAU Circular* no. 1212 (1949); "The Second Satellite of Neptune," PASP 61 (1949): 175–76; "The Fifth Satellite of Uranus," PASP 61 (1949): 129; "Some Results on Planets (abstract)," AJ 54 (1949): 191; "Planetary Atmospheres and Their Origin," pp. 306–404 in Kuiper, *Atmospheres of the Earth and Planets;* see also the previous version, "Survey of Planetary Atmospheres," in the 1949 edition, which preceded this intensive observational work.

69. See Carl Sagan, *Cosmos* (New York: Random House, 1980), p. 143 footnote; Sagan, "Obituary, Gerard P. Kuiper," *Icarus* 22 (1974): 117–18.

70. Kuiper, ed., *Atmospheres of the Earth and Planets.*

71. Brush, "Theories of the Origin of the Solar System," pp. 85–91.

72. Yerkes astronomer Thornton Page wrote: "The emphasis is shifted from the origin of the earth, as one of the planets, to the origin of the solar system as a whole. The latest trend goes even further in linking the origin of the solar system with the early history or origin of our galaxy of stars and even of the whole universe." Thornton Page, "The Origin of the Earth," *Physics Today* 1 (October 1948), reprinted in *Annual Report of the Board of Regents of the Smithsonian Institution,* 81st Congress, 2nd Session (Washington, D.C.: GPO, 1950).

73. See Harlan J. Smith, *Life in the Universe,* " Book Review in *Science* 220 (20 May 1983): 846–47: "Widespread acceptance of this cosmogony gave an entire generation, up to the 1940s, little reason to think very seriously about life elsewhere in the galaxy. . . . [various evidence] led, after the Second World War, to an almost euphoric acceptance by most astronomers that life could well be a ubiquitous phenomenon in the universe."

74. Daniel L. Harris, "Photometry and Colorimetry of Planets and Satellites," pp. 272–342 in Kuiper and Middlehurst, eds., *Planets and Satellites;* G. P. Kuiper, Y. Fujita, T. Gehrels, I. Groenevelt, J. Kent, G. van Biesbroek, and C. J. Van Houten, "Survey of Asteroids," *Astrophysical Journal,* Supplement, 3/32 (1958): 289–427 and G. P. Kuiper, "Limits of Completeness," pp. 575–91 in *Planets and Satellites;* Carl Sagan, "Physical Studies of Planets" (Ph.D. Thesis, University of Chicago, 1960).

75. G. P. Kuiper, D. W. G. Arthur, E. Moore, J. W. Tapscott, *Photographic Atlas of the Moon* (Chicago: University of Chicago Press, 1960).

76. *Transactions of the International Astronomical Union* VIII (1952): 208–17; IX (1955): 250–63; X (1958): 249–64.

77. In April 1947, while the 200-inch was being tested before going into full operation, Edwin Hubble used the canals of Mars as one example of problems that could be attacked with the instrument. Edwin Hubble, "The 200-Inch Hale Telescope and Some Problems It May Solve." Alexander F. Morrison Lecture, Pasadena, California, 8 April 1947, reprinted in *Annual Report of the Board of Regents of the Smithsonian Institution,* 81st Congress, 2nd Session (Washington, D.C.: GPO, 1950), pp. 175–88.

78. E. C. Slipher and A. G. Wilson, *Mars: 1954: Astronomical Society of the Pacific Leaflet 301* (June 1954); R. I. Mitchell, *The 1954 International Mars Photographic Patrol.* Report of the International Mars Committee (Flagstaff, Ariz: Lowell Observatory, 1955); "Reports of Observatories," *Astronomical Journal . . . ;* Audouin Dollfus, President, Commission 16 IAU, "Memorandum: Collective Observations in Some Programs of Study of the Planets" (1 July 1960), copy in WNRC 255-66A-1318: 6 ("Subcommittee Papers"); *Transactions of the International Astronomical Union, passim.*

79. Robert S. Richardson, "A Postwar Plan for Mars," *Astounding Science Fiction* (Jan. 1944), quoted in Robert S. Richardson (Mount Wilson and Palomar Observatories), *Exploring Mars* (New York: McGraw-Hill, 1954), pp. 19–20; Robert S. Richardson, *Some Observations Made of Mars at Mount Wilson in 1956: Astronomical Society of the Pacific Leaflet 333* (Feb. 1957).

80. Richardson, *Exploring Mars,* pp. 20–21. Richardson, at Mount Wilson, could have been referring to the rocket engineers at the Jet Propulsion Laboratory at CalTech, although he also would have had access to the opinions of the engineers at White Sands through discussions with Clyde Tombaugh on the International Mars Committee.

81. See Dollfus's memoir, "Pioneering Balloon Astronomy in France," *Sky and Telescope* 66 (Nov. 1983): 381–87.

82. Harold Urey, *The Planets: Their Origin and Development* (New Haven: Yale University Press, 1952), p. ix.

83. Carl Sagan, "Harold Clayton Urey—In Memoriam," *Icarus* 48 (1981): 348–52; Cyril Ponnamperuma, "Harold Clayton Urey: Chemist of the Cosmos," *Sky and Telescope* 61 (May 1981): 397; Brush, "Harold Urey and the Moon," op. cit.; Joseph N. Tatarewicz, "Harold Clayton Urey," *Dictionary of Scientific Biography II* (forthcoming).

84. Urey, *The Planets.* Harrison Brown had contributed to the Yerkes 50th anniversary symposium which Kuiper had arranged in 1947, speaking on the evolution of the earth's atmosphere as a secondary product of the formation of the earth, instead of, as previously believed, a remnant of the solar nebula. Brown, Urey, and others at the Institute for Nuclear Studies at Chicago were engaged in using the new techniques and equipment which had developed out of the Manhattan Project to analyze and date rocks and meteorites.

85. Brush, "From Bump to Clump"; Cruikshank, "20th Century Astronomer," p. 163.

TWO. DEFINING THE NEED (1958–1963)

1. Walter McDougall, *The Heavens and the Earth: A Political History of the Space Age* (New York: Basic Books, 1985); David H. DeVorkin, *Race to the Stratosphere: Manned Scientific Ballooning in America* (New York: Springer Verlag, 1989); David H. DeVorkin, *Science with a Vengeance: The Origins of Space Science in the V-2 Era* (New York: Springer Verlag, forthcoming).

2. R. Cargill Hall, *Lunar Impact: A History of Project Ranger* (Washington, D.C.: NASA SP-4210, 1977), pp. 12–13.

3. Ibid., p. 13.

4. This followed four unsuccessful attempts by the Air Force and Army to send Pioneer probes past the moon, although a fifth attempt did pass about 60,000 km from the moon in March 1959; see Hall, *Lunar Impact*, pp. 6–10.

5. Robert Jastrow, "Exploring the Moon," pp. 45–50 in Paul A. Hanle and Von Del Chamberlain, eds., *Space Science Comes of Age: Perspectives in the History of the Space Sciences* (Washington, D.C.: Smithsonian Institution Press, 1981); Brush, "Nickel For Your Thoughts," 891–98, and "Harold Urey and the Moon: The Interaction of Science and the Apollo Program," pp. 430–70 in Peter M. Bainum and Dietrich E. Koelle, eds., *Spacelab, Space Platforms, and the Future: Proceedings of the Twentieth Goddard Memorial Symposium* (San Diego, Calif.: American Astronautical Society, 1982).

6. Homer E. Newell, "Harold Urey and the Moon," *The Moon* 7/1–2 (March/April 1972): 1–5; Newell, "Paper on Harold Urey and the Moon for Kopal's Journal," typescript attached to memo to distribution, 6 Sept. 1972, WNRC 255-79-0649 (26): "AA READING FILE"; Newell, memo to file, Telephone Conference Report with Robert Cowen, *Christian Science Monitor*, 6 Jan. 1965, JPLHF 2–730.

7. Newell to Silverstein, "Recommendations for a Lunar Science Group," op. cit.

8. NASA News Release, "NASA Will Formulate Space Science Working Groups," 20 Mar. 1959; Herbert S. Fuhrman to Newell, 16 Feb. 1959; "Office of Space Sciences Program Planning," 28 April 1959; all in WNRC 255-79-0649: 11 (170 and 171), "READING FILE"; OSS/OSSA Staff Meeting Minutes, WNRC 255-79-0649: 2 (48), passim; see especially Newell to Silverstein, "Proposed Functional Responsibilities and Personnel Complement for Office of Space Sciences," 30 July 1959, and attached organization chart.

9. OSS/OSSA Staff Meeting Minutes, 21 July, 28 July, 30 July, and 21 Aug. 1959. The discipline breakdown was: atmospheres, ionospheres, energetic particles, electric fields, magnetic fields, planetology, astronomy, biosciences, and special problems. Astronomy, under Nancy Roman, included solar physics; planetology, under chemist Robert F. Fellows and geophysicist John O'Keefe, included surfaces and interiors of bodies. In place of the ad hoc panels of consultants, a regular system was instituted in April 1960, with a Space Sciences Steering Committee supported by subcommittees formed generally according to the above disciplinary breakdown. The rationale for various divisions of the space sciences area is also discussed in "Office of Space Sciences Ten Year Program," 17 Aug. 1959, pp. 3–4 and section III (NHOA).

10. When Nancy Roman arrived in March 1959 to assume responsibility for observational astronomy, she was told that "astronomy is the study of where you aren't." The lunar and planetary research was seen as an extension of geology and geophysics, and was carried out in another division of the Office of Space Sciences. Roman recalled little interaction between the astronomy (earth orbital) and lunar and planetary (deep space probe) sides. Nancy G. Roman, OHI by Tatarewicz, 28 Jan. 1983 (SAOHP), pp. 2–3.

11. For discussions of the significance of organizational structure, and the major outlines of the overall management of the agency see Robert Rosholt, *An Administra-*

tive History of NASA, 1958–1963 (Washington, D.C.: NASA SP-4101, 1966) and Arnold S. Levine, *Managing NASA in the Apollo Era* (Washington, D.C.: NASA SP-4102, 1983). On the vehicle and discipline breakdowns see Newell, *Beyond the Atmosphere.*

12. "Ten Year Program," op. cit, pp. 4–5 and section 5, and other planning documents in NHOA. This was also in keeping with Administrator Glennan's reluctance to consider planetary probes except in the context of future planning. See Newell, *Beyond the Atmosphere*, pp. 262–63; Ezell and Ezell, *On Mars*, 1, pp. 15–20; Hall, *Lunar Impact*, pp. 16–24.

13. Ezell and Ezell, *On Mars*, pp. 31–32.

14. Ibid., pp 26–29.

15. Albert R. Hibbs, OHI by Allan A. Needell and Joseph N. Tatarewicz, 9 Dec. 1982 (SAOHP), p. 14, pp. 4–8; Ezell and Ezell, *On Mars*, pp. 16–17; Hall, *Lunar Impact*, p. 17.

16. Hibbs, OHI 8 Dec. 1982, p. 7; Hibbs, OHI by R. C. Hall, 2 Oct. 1972, JPLHF 3-595, pp. 31–32.

17. Newburn, OHI, pp. 39–46.

18. Ezell and Ezell, *On Mars*, pp. 16–20; Hall, *Lunar Impact*, pp. 45–49; Clayton R. Koppes, *JPL, and the American Space Program* (New Haven: Yale University Press, 1982) pp. 99–103; Hibbs interview by Hall, pp. 9–11, 14; copies of the Juno IV proposals, and a copy of R. Newburn and M. Neugebauer, "Preliminary Consideration of a Limited Class of Problems Suitable for Study by Interplanetary Probes and/or Satellites," 13 Nov. 1958, in Newburn, SAOHPWF.

19. A. R. Hibbs, ed., "Exploration of the Moon, Planets, and Interplanetary Space" (Pasadena, Calif.: Jet Propulsion Laboratory Technical Report 30-1, 30 April 1959).

20. Ibid, pp. 88–90. The statements concerning the need for ground-based planetary astronomy do not appear in the earlier (November) draft. However, the draft had been circulated to various astronomers and others known to be sympathetic to planetary research, and Newburn, Neugebauer, and others had interviewed a number of such astronomers. Also, Hibbs had been one of the organizers of the Lunar and Planetary Exploration Colloquium, where discussion of the need for more planetary astronomy had been frequent. Finally, JPL planners had been in touch with, and had received a number of documents from, the Space Science Board of the National Academy of Sciences, in which similar sentiments were expressed. Hugh Odishaw, Executive Director Space Science Board to W. H. Pickering, JPL, 5 Feb. 1959 with listing of SSB documents sent, WNRC 255-79-0649: 11 (163) "JPL"; the Space Science Board discussions of needed planetary astronomy will be treated below; *Proceedings of the Lunar and Planetary Exploration Colloquium* (Downey, Calif.: North American Aviation Missile Division, 1958–63), hereafter cited as LPEC.; see especially the attendance lists and the proceedings of the following meetings: 13 May 1958 (1); 15 July 1958 (2); 29 October 1958 (3); 12 January 1959 (4); 18 March 1959 (4).

21. Hibbs, OHI, pp. 11–15, and letter to the author 2 May 1983. Note the similarity to the experience of radio astronomers in Britain, who as outsiders to astronomy started to learn the fundamentals by reading popular books and listening to programs on the radio: David Edge and Michael Mulkay, *Astronomy Transformed: The Emergence of Radio Astronomy in Britain* (New York: Wiley, 1978).

22. John E. Froelich, memo, "Minutes of Meeting on N.A.S.A. Space Program of October 27 [and 28], 1958," 28 [and 29] October 1958, JPL Library Archives, quoted in Ezell and Ezell, *On Mars*, pp. 17–18; see also Koppes, *JPL.*

23. Pickering, "Edited Notes on Phone Call to [Associate Administrator Richard E.] Horner in NASA," 8 Dec. 1959," JPLVC Roll 311-64, quoted in Koppes, *JPL,* p. 105.

24. Horner to Pickering, 16 Dec. 1959 WNRC 255-79-0649: 11 (163) "JPL".

25. Newell, memo to file, "Trip Report for the Visit to the Jet Propulsion Laboratory on 28 December 1959 by Homer E. Newell, Jr., Newell Sanders, J. A. Crocker, Morton J. Stoller," 30 Dec. 1959, pp. 1–3, in "JPL" op. cit.

26. Newell, "Trip Report," pp. 6–7; Morton J. Stoller, "Trip Report on Visit to JPL, 28–29 December 1959," 12 Jan. 1960, in "JPL," op. cit; Abe Silverstein to Pickering, 26 Jan. 1960. The latter sets forth the membership and charter of the NASA Steering Group on Lunar, Planetary, and Interplanetary Exploration and the NASA Committee on Lunar, Planetary, and Interplanetary Science. These were quickly superseded by an overhaul of the entire advisory apparatus in April, which gathered all of the various ad hoc discipline subcommittees into a Space Sciences Steering Committee. See below.

27. In the parlance of recent approaches in the social studies of science and technology, the JPL and other proponents of planetary astronomy were engaging in heterogeneous engineering, building actor networks, or building systems. See chapter 5 below for a fuller analysis.

28. J. G. Beerer, President, Missile Division, North American Aviation to Homer E. Newell, 20 Oct. 1960 WNRC 255-79-0649: 13 (191) "R & A"; LPEC.

29. LPEC 1, "Preface" (May 1960).

30. A. G. Wilson, "Syntax of Space Exploration," LPEC I/2 (15 July 1958): 16–18. The development of astronomy and planetary science at the RAND Corporation, an institution primarily concerned with problems of nuclear strategy, is the subject of a current study by the author.

31. LPEC I/5 (18 March 1959): pp. 22–27, 28.

32. S. M. Greenfield, "Introduction," LPEC II/1 (Sept. 23–Apr. 19 1959: 1).

33. LPEC II/1 (23–24 September 1959): 34–35.

34. Recapitulation and Appraisal," LPEC, pp. 52–53.

35. JPL did just this. See Hibbs's remarks in LPEC I/5 (18 March 1959): 28, and G. A. Derbyshire memo for the record, "Telephone Conversation with Richard Davies, JPL," 29 March 1960, in NAS:ADM:ORG:NAS:SSB:Committees: Optical & Radio Astronomy: General. JPL's aggressive recruiting campaign for astronomical expertise, the high salaries they offered for junior astronomers, and their reputation in segments of the scientific community would continue to cause them problems. See: Newell, *Beyond the Atmosphere,* pp. 266–67; Koppes, *JPL,* p. 213. See also Nancy G. Roman to John Clark, 17 May 1963, "Relations between Jet Propulsion Laboratory and the Astronomical Community," WNRC 225-74-663 (49) "SGA Reading File."

36. Ibid., p. 54. Many meteorological changes take place on Mars in rapid time scales, but since both Mars and the Earth rotate at about the same rate, features can only be observed continuously if the planet is picked up by a second observatory immediately west of the first.

37. Ibid., p. 55. See also Herrick's comments on p. 38 where he points out that the *American Ephemeris* does not always use the *latest* values, but attempts to maintain some consistency so that long series of observations can be referred to a standard system. The engineers, on the other hand, were accustomed to finding the latest values for various data in handbooks.

38. Greenfield, ibid., p. 57.

39. Newell, *Beyond the Atmosphere,* p. 205, and especially chapters 5, 12, and appendix F, Membership of the Space Science Board, 1958–1972; Norriss S. Hetherington, "Winning the Initiative: NASA and the U.S. Space Science Program," *Prologue* (Summer 1975): 99–107; Hetherington, "The Evolution of NASA's Science Program, 1958–1960," NASA Historical Note HHN-123, August 1972, NHOA; Charles M. Atkins, "NASA and the Space Science Board of the National Academy of Sciences," NASA Historical Note HHN-62, Sept. 1966, NHOA.

40. ADM:ORG:NAS:SSB:Committees: Optical and Radio Astronomy: Report: Interim; other members of the committee were: Otto Struve, Martin Schwarzschild,

Lyman Spitzer, Horace Babcock, Arthur Code, Frederick T. Haddock, John Findlay, and Roger Gallet.

41. NAS:SSB:Group on Physics of Planetary Surfaces and Atmospheres Jt w Armed Forces-NRC Com on Bio-Astronautics:Proposed; see also ORG:NAS:SSB: Committees: Chemistry of Space and Exploration of the Moon and Planets:General; also Report File:ADM:ORG:SSB:Committees:WESTEX:AdHoc:Summary Report: References:No. 15. Replies to Dyer's memo and enclosures are found in all of the above folders.

42. See below.

43. See below.

44. These various positions can be found in Horace W. Babcock to Leo Goldberg, 31 May 1960, in NAS:ADM:ORG:NAS:SSB:Committees:Optical and Radio Astronomy:General, as well as NAS:ADM:ORG:NAS:SSB:Committees:Chemistry of Space and Exploration of the Moon and Planets:Report:Interim, p. 1. Membership of the committee included Harrison Brown, Harmon Craig, and Mark Inghram (all alumni of the University of Chicago and colleagues of Urey and Kuiper in the early 1950s), Frank Press (Caltech), Gerard de Vaucouleurs, and Fred Whipple.

45. SSB Minutes of the Fifth Meeting, 5–7 May 1959, pp. 7–9, copy in NHOA. On Urey's reaction to JPL's Report 30-1 and Newburn's contribution see: Urey to Newburn 3 July 1959, 14 July 1959, Urey to Hibbs 10 Aug. 1959, all in Ray L. Newburn, SAOHPWF.

46. Berkner to Kistiakowski, 13 Nov. 1959, with attached "Statement of L. V. Berkner, Chairman, Space Science Board, National Academy of Sciences, to the President's Science Advisory Committee Concerning The Space Science Program of the United States," pp. 7–8, copy in NHOA.

47. Interim Report no. 2, 9 Mar. 1960.

48. Shapley to members of Committee 7, SSB, 6 Oct. 1959, "Meeting with Dr. Harold Urey . . ."; Shapley to Urey, 6 Oct. 1959; Anon., memo for files, "Translation of Shapley's Notes on Discussions with Urey re Planetary Atmospheres," 28 Mar. 1960; Report on Committee on Ionospheres of the Earth and Planets, 8 Mar. 1960; all in NAS:ADM:ORG:NAS:SSB:Committees:Ionospheres of the Earth and Planets:Meetings:Chairman Jt w Chairman SSB Com. on Chemistry of Space and Exploration of Moon and Planets.

49. Space Science Board, Minutes of Eighth Meeting, Saturday, 25 June 1960, pp. 6–7. The proceedings of the conference on planetary atmospheres was prepared by Carl Sagan and William W. Kellogg from stenographic transcripts and then edited in several cycles, including additional meetings at JPL of an ad hoc Panel on Planetary Atmospheres on December 15–17, 1960 and February 2–4, 1961. The proceedings appeared as *The Atmospheres of Mars and Venus,* A Report by the Ad Hoc Panel on Planetary Atmospheres of the Space Science Board, prepared by William W. Kellogg and Carl Sagan (Washington, D.C.: NAS Publication 944, 1961); but see NAS:ADM:C&B:SSB:Panels:Planetary Atmospheres:AdHoc: 1961–1962, ORG:NAS:SSB:Conferences:Planetary Atmospheres:Arcadia (Calif.): Proceedings.

50. Newell, *Atmosphere,* p. 205, also pp. 124, 212; Homer E. Newell, "Workpaper #8—The Extent and Adequacy of Relationships between NASA and the Scientific Community," 17 June 1960, WNRC 255-79-0649: 11 (161) "Advisory Committee Luncheon 23 June 1960 Occidental Restaurant."

51. L. A. Manning to Alan H. Shapley, 28 June 1960, sent as Space Science Board Memorandum SSB-167, 7 July 1960. Newell's copy is in the NASA History Office Archives, "Space Science Board—1960."

52. Memorandum, Kellogg to Berkner, 6 Feb. 1961, NAS:ADM:C&B:SSB: Committees:Upper Atmosphere Rocket Research:Report, and NAS:SSB:Panels: Planetary Atmospheres:Adhoc:Meetings:Report. See also Lederberg to Berkner and others, 5 June 1961 with attached resolutions and copy of Lederberg to Johnson,

1 May 1961 in NAS:ADM:C&B:SSB:Committees:Chemistry of Space and Exploration of the Moon and Planets: General.

53. Berkner to Webb, 31 Mar. 1961, enclosing "Support of Basic Research
for Space Science," 27 Mar. 1961 NAS:ADM:C&B:SSB:Meetings:9th:1961 Feb.

54. NASA Management Instruction (NMI) 37-1-1, 15 Apr. 1960, "Establishment
and Conduct of Space Sciences Program—Selection of Scientific Experiments,"
and NASA Circular No. 73, 37-1-2, May 27, 1960, "Membership of Space Sciences
Steering Committee and Subcommittees," copies in NHOA and SAOHPWF.

55. Summary Minutes, Astronomy Subcommittee of the Space Sciences Steering
Committee (mtg. no. 1), 25 Apr. 1960. Unless otherwise indicated, all citations
will be to these official minutes. Copies of minutes, originals or copies of correspondence, and additional handwritten notes by Newburn are in SAOHPWF.

56. Minutes, pp. 5–6. Kuiper's memorandum is discussed below.

57. Minutes, p. 6.

58. Minutes, 7 June 1960, p. 2. The subcommittee minutes and correspondence
may be found in the same places as those of the Astronomy Subcommittee. The
Planetary and Interplanetary Sciences Subcommittee will hereafter be referred to
as the "P & I Subcommittee."

59. Minutes, p. 8. Kuiper had drafted an eleven-page memorandum, "Need
for a Ground-Based Lunar and Planetary Observatory," June 1960. This memorandum was prepared for the NASA subcommittee, but it is unclear whether the
subcommittee members were given copies at this meeting, or whether Kuiper was
asked to write up his presentation for consideration.

60. Newell to Derbyshire, SSB, June 1960, with copy of Kuiper's memorandum
and replies from various reviewers attached, WNRC 255-79-0649: 6 (103) "Space
Science Board-Material Furnished for Information"; James K. Gleim (Secretary,
Astronomy Subcommittee) to Newburn, 7/14/60, SAOHPWF, where Newburn (and
other members of the Astronomy Subcommittee) are requested to discuss the memo
at the July meeting.

61. Kuiper, "Need," pp. 1–2.

62. Kuiper, "Need," pp. 3–5.

63. Kuiper, "Need," p. 5, and pp. 5–8.

64. Kuiper, "Need," p. 8.

65. Kuiper, "Need," p. 9.

66. Kuiper, "Need," p. 10.

67. Kuiper to Newell, 28 June 1960 with attached Greenstein to Kuiper, 15
June 1960, WNRC 255-66A-1318: 6 (no number) "Subcommittee Papers." Copies
were sent to E. M. Cortright, Director of Lunar and Planetary Programs, Roger
C. Moore, Lunar and Planetary Sciences, and Nancy Roman.

68. Replies were received from Urey, Stuhlinger, Philip Abelson, Edward P.
Ney, Joseph W. Chamberlain, Fred Whipple, and others. These replies are attached
to Newell's copy of the Kuiper memorandum.

69. Mulders to Newell, 1 Aug. 1960. The Air Force attempt to establish a
planetary observatory at Cloudcroft, New Mexico will be discussed below. At
this point it should be noted that a year earlier NASA had to cancel the JPL
Vega launch vehicle because of a sudden disclosure that the Air Force had secretly
been working on an uprated Agena B of similar capabilities, and NASA was under
pressure not to compete with or duplicate expensive facilities which were under
construction elsewhere.

70. Nancy Roman, "Trip Report—Visit to Sacramento Peak Observatory—21
May 1960," 2 June 1960, WNRC 255-66A-1318: 6 (no number) "Subcommittee Papers."

71. In September 1960, Newell wrote: "With the availability of rockets and
satellites, the opportunity is opening up for the scientist to study the planets from

much the same viewpoint as that from which he has been studying the earth in the past. Once the property of the astronomer, and not too highly valued a property at that, the planets are now brought within the purview of the geophysicist. It is hoped that geophysicists and astronomers will now unite to give them the attention that they deserve scientifically." H. E. Newell, R. Jastrow, G. MacDonald, "A Home for Planetary Sciences," prepared for the President's Page in *Transactions of the American Geophysical Union,* Draft MS (June 1960) in WNRC 255-79-0649: 4 (66) folder, "AGU Planning Committee on Planetary Science, 1960," published in *Transactions of the American Geophysical Union* 41 (Sept. 1960): 407–409.

In July 1964, Newell showed a slide of the solar system to a group of educators and said "the domain shown in this last slide is the new kingdom of the geophysicist." Newell, draft of "The Impact of Space Techniques on Modern Science," 6 July 1964, WNRC 255-79-0649: 12 (185) "Chron File . . . ," p. 21.

72. Minutes, 7 Sept. 1960, pp. 7–8.

73. Ernest J. Ott to Ray L. Newburn, 20 Dec. 1960 and Newburn to Ott, 10 Jan. 1961.

74. Minutes, 23–24 Feb. 1961, p. 7; "Summary of Comments Received from the Astronomy Subcommittee Suggesting Needed Planetary Research," [prepared by Ernest J. Ott] for meeting no. 5 of the P & I Subcommittee, 23–24 Feb. 1961. Copy in WNRC 255-66A-1318: 6 (no number, miscellaneous papers loose in box).

75. Newburn solicited comments on needed planetary research from the members of his "sub-subcommittee." See O'Keefe to Newburn, 23 Feb. 1961; Newburn to Brouwer, 10 Mar. 1961; Newburn to Jones, 10 Mar. 1961; Brouwer to Newburn, 11 Mar. 1961; all in SAOHPWF; Minutes, meeting no. 6, Astronomy Subcommittee, 22–25 Mar. 1961; Newburn to Roman, transmitting sub-subcommittee findings, 28 Mar. 1961; Newburn to Roman, 11 Apr. 1961; various correspondence concerning revisions to meeting notes, passim; all in Newburn, SAOHPWF.

76. Whipple to Edmondson, 18 Oct. 1960, quoted in J. A. Hynek, ed., "A Recommendation for a Ground-Based and Balloon-Borne Lunar-Planetary Observation Program in Support of the United States Program of Space Exploration" (Pasadena, Calif.: Jet Propulsion Laboratory Technical Memorandum 33–37, revised 1 Mar. 1961), p. 2.

77. Ibid., pp. 1–3.

78. Ibid., pp. 4–5.

79. Ibid., pp. 5–7. Reports of changes on the surface of the moon were made from time to time, but such reports were seldom believed. In 1959 the Soviet astronomer Nikolai Kozyrev of the Pulkovo Observatory obtained a spectrum apparently showing some sort of gaseous emission from the crater Alphonsus. Since then, Lunar Transient Phenomena, LTPs as they came to be called, have been cataloged and investigated by NASA and others. This proposed patrol of the lunar surface for LTPs was subsequently adopted, as were selected other recommendations of the report, and will be discussed below.

80. Ibid., pp. 7–8, 8–11, 15–18.

81. Ibid., pp. 22–23.

82. Nancy Roman to Roger C. Moore, Head, Planetary Sciences, [through Assistant Directors for Satellite and Sounding Rocket Programs, and Lunar and Planetary Sciences, respectively], "JPL Technical Memorandum No. 33–37—A Recommendation for a Ground-Based and Balloon-Borne Lunar and Planetary Observation Program," 23 Oct. 1961 WNRC 255-74-663 (48) "Dec. 1959–1961 Astronomy and Solar Physics."

83. Ibid., pp. 1–2.

84. Ibid., pp. 6, 5.

85. Newell, in a quick review for Administrator Fletcher in 1971, recalled, "In the first year of NASA's existence, the space science group in Headquarters

recommended providing some support to ground-based astronomy because of its potential role in laying the groundwork for future space missions and because of the foreseeable need to have ground-based observations conducted simultaneously with space-borne solar and planetary missions. At that time, [Administrator] Dr. Dryden felt that NASA could not justify giving direct support to ground-based astronomy, so he ruled against the OSSA plans." Newell to Fletcher, 28 Oct. 1971, "OSSA Astronomy Program," WNRC 255-79-0649 (25): "AA Reading Files."

86. Barbara Bell, ed., "Research Study to Consider the Establishment of an Observatory for Planetary Studies near Cloudcroft, New Mexico" (Harvard University, Donald H. Menzel, Principal Investigator, AFOSR contract AF49(638)-1155, final report 30 Apr. 1962), p. 3, hereafter referred to as the "Menzel Report." A copy of this report and a selection of correspondence concerning the Cloudcroft facility was provided by one of the committee members, Dr. Frank K. Edmondson, for which the author is grateful. Materials so provided will be identified by "FKE."

87. Bradford A. Smith and John W. Salisbury, *Final Report on the Evaluation of the Cloudcroft, New Mexico Site for a United States Air Force Planetary Observatory* (University Park: New Mexico State University Research Center/Geophysics Research Directorate, Air Force Cambridge Research Laboratories Contract AF 19(604)-7443, Report 1064, December 1961, revised 19 Mar. 1962). This report, of which there are two versions, will be cited as "Report 1064 (1961)" or ". . . (1962)." See also Menzel Report, p. 7.

88. Report 1064 (1961), pp. 80, 98. The procedures used to evaluate the seeing at Cloudcroft and at Kitt Peak are discussed at length in the Menzel Report, pp. 7–13, which constitutes a transcript of the deliberations of the committee and discussions with Salisbury, Tombaugh, and Smith.

89. Menzel Report, pp. 1–2. The committee members, in addition to Menzel, were: Ira S. Bowen (Director, Mount Wilson and Palomar), Frank K. Edmondson (Indiana University), John S. Hall (Lowell), Nicholas U. Mayall (Director, Kitt Peak), Aden B. Meinel (Arizona), C. D. C. Shane (Lick). Lewis Kaplan, JPL, "also agreed to come to represent the important studies being conducted by NASA and JPL in the lunar-planetary field."

90. Menzel Report, p. i, and deliberations, passim; Barbara Bell to Menzel Committee members, 26 Mar. 1962 (FKE); Report 1064 (1961), pp. iii, 80, 98; compare with Report 1064 (1962), pp. iii, 80, 98.

91. Menzel Report, p. 3.

92. Menzel Report, pp. 8, 13.

93. Menzel Report, pp. 21–22. The research program of Salisbury's Lunar and Planetary Exploration Branch as it existed in 1962 is summarized in appendix A of the Menzel Report.

94. A 48-inch reflector was eventually installed at Cloudcroft and became operational in 1964. An unusual instrument, in that it was on an equatorial mount which itself was on an azimuth mount, the reflector was designed to photograph (on film and on image tubes) satellites in orbit. The telescope had the capability for visual and photographic observations, laser tracking, photometry and radiometry in the 10-micron band, and was computer controlled. As of 1968, observers were able to observe paint patterns on a booster at 160 miles altitude "gradually gray out in about two weeks due to ultraviolet erosion of the paint." The space track operations were later moved to sites elsewhere, and Sacramento Peak took over the telescope for a varied program of stellar photometry and some planetary photography. See Frank A. Burnham, "Cloudcroft Pressing Observation Methods," *Aerospace Technology* (26 Feb. 1968): 30–31; Joel W. Powell, "Photography of Orbiting Satellites," *Spaceflight* 25 (2 Feb. 1983): 82–83; and Timothy Schneeberger, et al., "Cloudcroft Observatory Today," *Sky and Telescope* 59 (Feb 1980): 109–110.

95. Ostrander to ?, 7 May 1964 (unable to locate); Roman to Salisbury, ca.

Aug. 1963 (unable to locate); John Naugle, Director of Geophysics and Astronomy to Newell (drafted by Roman), 16 July 1964; Newell to Dryden, 2 Oct. 1964, "Planetary Astronomy—The Critical Need for Telescopes," WNRC 255-67A-601: 7 "Astronomy 1–2."

96. Menzel Report, p. 5.

97. Menzel Report, pp. 5, 7, and appendix A.

98. The divergent viewpoints between JPL and NASA headquarters appear with this issue as well. Roman to Naugle, 17 May 1963, op. cit.; Newburn, OHI, esp. pp. 70–81; William E. Brunk to file, 18 Dec. 1964, "Trip Report," AR. Al Hibbs, chief of space sciences at JPL at the time, later recalled, "As far as the astronomers represented by Caltech, they wanted nothing to do with us. They were astrophysicists, and they thought we were taking money that should go to astrophysics to waste our time looking at the planets, and the moon, an uninteresting object. So they were a little bit upset by the whole program and made it fairly clear." Hibbs, OHI, p. 16, see also Hibbs to Tatarewicz, 2 May 1983, in SAOHPWF.

99. Newburn, OHI, pp. 105–110. Newburn, "Annual Report on Progress in Astronomy for the Astronomical Journal," typescript, ca. Aug. 1962; Newburn, "For the Annual 'Reports of Observatories' in the Astronomical Journal," typescript, ca. Aug. 1963; Newburn, "Contribution to the Astronomical Journal," typescript, ca. Aug. 1964, all in SAOHPWF.

100. Attempts were made in 1963–64 to obtain a 60-inch telescope for Caltech which could be used by the JPL astronomers, but the plans fell through in the final stages. A joint effort by JPL and the University of Washington in 1967 to get an 84-inch telescope on Table Mountain also fell through. This will be discussed below, as will the successful acquisition of an unusual 24-inch telescope for Table Mountain.

101. G. P. Kuiper, "Organization and Programs of the Laboratory," *Communications of the Lunar and Planetary Laboratory* 1 (12 Feb. 1962): 1, hereafter cited as CLPL. See also Kuiper, "The Lunar and Planetary Laboratory," *Sky and Telescope* (Jan. 1964): 4–7 and (Feb. 1964): 88–92. Ewen A. Whitaker, *The University of Arizona's Lunar and Planetary Laboratory: Its Founding and Early Years* (Tucson: University of Arizona, 1986).

102. CLPL 1, pp. 1–2, 20.

103. Ibid., passim; "The Lunar and Planetary Laboratory and Its Telescopes," CLPL 172 (31 Dec. 1972); Nancy Roman to file, "Visit to the University of Arizona, November 1, 1962," 21 Nov. 1962, WNRC 255-74-663: 48 "Reading File, Jan.–June, 1964". Aden Meinel had been at Yerkes from 1950 to 1956, when he became the first director of Kitt Peak during its planning and construction phase, 1956–60. He then moved to Arizona to become Astronomy Department chairman. An expert and ambitious optical designer, Meinel had worked on designs for large orbiting telescope mirrors. Harold Johnson had developed the U, B, V, etc. standard system of stellar photometry, and was an expert in optical and electrical systems design. He had been at the Harvard Radiation Laboratory during the war, and afterward spent two years at Yerkes, seven years at Lowell, and three years at McDonald.

104. Hall, *Lunar Impact,* p. 79; Roger Moore to Newell, 6 Dec. 1961, WNRC 255-67A601: 21, "SL Chron." See also Roman to file, 21 Nov. 1962, op. cit.

105. Compare CLPL 1 (op. cit.) with Kuiper, "The Lunar and Planetary Laboratory and Its Telescopes," CLPL 172 (31 Dec. 1972): 199–247) for a summary description of what Kuiper and his collaborators were able to do in just a decade. On the concern over Kuiper's many projects see: William E. Brunk, Untitled Draft MS, 27 Oct. 1967, AR, Nancy G. Roman, OHI. Kuiper drew fire from his professional colleagues for establishing the CLPL; Newell, *Beyond the Atmosphere,* pp. 127–28.

THREE. FUNDS AND A FOCUS (1963–1970)

1. See chapter 5 for aggregate funding information and demographic analysis.

2. Rosholt, *Administrative History of NASA* (Washington, D.C.: NASA SP-4101, 1966), pp. 221–25; Arnold S. Levine, *Managing NASA in the Apollo Era* (Washington, D.C.: NASA SP-4102, 1983), pp. 36, 41–42.

3. Newell, *Beyond the Atmosphere,* pp. 207–212.

4. National Academy of Sciences, Space Science Board, *A Review of Space Research.* (Washington, D.C.: NAS, 1962), pp. 2–26, 4–6. The study of ground-based planetary astronomy was eventually conducted by a panel chaired by John S. Hall of Lowell Observatory and its report appeared as: National Academy of Sciences, Space Science Board, Panel on Planetary Astronomy, *Planetary Astronomy: An Appraisal of Ground-Based Opportunities.* (Washington, D.C.: NAS, 1968). This will be discussed in chapter 5.

5. Pp. 2–26.

6. F. Clark, draft of "NASA Comments on 'A Review of Space Research,'" 15 May 1963, WNRC 255-79-0649: 5 (92) "Space Science Board. General. 1963," pp. 14–15.

7. A. Derbyshire to Newell, 8 Jan. 1963, with copies of Kellogg to Hess, 5 Dec. 1962 and Hess to Kellogg, 8 Jan. 1963, WNRC 255-79-0649: 5 (92) "Space Science Board: General: 1963."

8. W. Nicks, interview by Tatarewicz, 2 Aug. 1983; Nicks to Distribution, 3 May 1962 WNRC 255-74-663 : 50 (SL Chron May 1962). Moore had been coordinating this aspect of the Lunar and Planetary Programs office with Nancy Roman in the Geophysics and Astronomy office for some time prior to the official designation.

9. To W. W. Morgan (McDonald), 8 June 1962; George W. Preston (Lick), 7 Nov. 1962; T. Haddock (Michigan), 7 Nov. 1962; E. G. Bowen (CSIR, Sydney, Australia), 7 Nov. 1962; A. W. Rodgers (Mt. Stromlo, Australia), 7 Nov. 1962; Theodore Dunham (Mt. Stromlo), 7 Nov. 1962; Hideo Hirose (Tokyo), 7 Nov. 1962; all in WNRC 255-74-663 : 50 "SL Chron".

10. A. Schorn, OHI by Tatarewicz, 27 July 1983, SAOHP.

11. Gerard De Vaucouleurs, *Physics of the Planet Mars: An Introduction to Aerophysics* (New York: Macmillan, 1954). One interesting aspect of the consensus on the Martian surface pressure is that the methods used, while sensitive to various ad hoc assumptions, all converged on the same pressure.

12. Lewis D. Kaplan, Guido Munch, and Hyron Spinrad, "An Analysis of the Spectrum of Mars, *Astrophysical Journal* 139 (1 Jan. 1964): 1–15. See also D. G. Rea, "The Atmosphere and Surface of Mars—A Selective Review"; Donald M. Hunten, "CO_2 Bands and the Martian Surface Pressure"; Guido Munch, "Summary Remarks on Mars"; all in Harrison Brown, et al., eds., *Proceedings of the Caltech–JPL Lunar and Planetary Conference, September 13–18, 1965* (Pasadena, Calif.: JPL Technical Memorandum 33–266, 15 June 1966).

13. Minutes, Astronomy Subcommittee (Meeting No. 1-FY64), 8–9 Aug. 1963, pp. 10–13. Liddel had been visiting the various subcommittees related to the lunar and planetary discipline for some time. In February 1963 he had been at a planetary atmospheres subcommittee meeting where W. W. Kellogg and others discussed the need for lunar and planetary astronomy from the ground. Liddel had said at that time that the lunar and planetary programs office was "looking into funding additional specific research in planetary atmospheres at some of the larger observatories." "Summary Minutes, Planetary Atmospheres Subcommittee," (Meeting No. 2–63), 7–8 Feb. 1963, WNRC 255-67A601: 21.

14. Ezell and Ezell, *On Mars,* p. 93; U.S. Congress, Senate, Committee on Aeronautical and Space Sciences, *Hearings on 1965 NASA Authorization,* pt. 1,

88th Congr., 2nd sess. (Washington, D.C.: 1964), pp. 131–34; Nicks to Newell, 9 Aug. 1963, "Changes to AVCO & GE Voyager Studies," WNRC 255-74-663: 50 "Chron. Aug.–Dec. 63"; D. W. Tomayko to Administrative Officer, 8 Oct. 1963, "Biweekly Activity Report," ibid. Present at the conference were, among others, L. D. Kaplan, G. P. Kuiper, W. M. Sinton, A. Dollfus, and C. C. Keiss.

15. Newell to Seamans, 2 Oct. 1964, "Planetary Astronomy—The Critical Need for Telescopes," WNRC 255-67A601: 7 "Astronomy."

16. Ezell and Ezell, *On Mars,* pp. 96–100.

17. Schorn, OHI, 27 July 1983. The main improvement used by Spinrad in taking the spectrogram was an improved photographic emulsion sensitized for the infrared.

18. Ezell and Ezell, *On Mars,* pp. 100–101.

19. See chapter 5, below. Using the systems interpretation one would consider NASA's planetary exploration program as a system uniting scientific, technological, and political actors and forces. A "reverse salient" is a portion of the system whose development lags, threatening to hold back all of it. On the concept of "reverse salient" and "critical problem" see Thomas Parke Hughes, "The Evolution of Large Technological Systems," pp. 51–82 in Wiebe E. Bijker, Thomas P. Hughes, and Trevor Pinch, eds., *The Social Construction of Technological Systems: New Directions in the Sociology and History of Technology* (Cambridge, Mass.: MIT Press, 1987), and Hughes, "The Seamless Web: Technology, Science, Etcetera, Etcetera," *Social Studies of Science* 16 (1986): 281–92.

20. Schorn, OHI, 27 July 1983. On the money being contributed by various lunar and planetary program offices see Newell to Semans, 2 Oct. 1964, op. cit., p. 3, where he notes that all funds for the telescopes "have been provided by the program Offices with OSSA as justified in direct support of their missions."

21. For the motivations and growth of the university program see Newell, *Atmosphere,* chapter 13; on Webb's view see especially pp. 233–36, and Webb's foreword to Arnold S. Levine, *Managing NASA in the Apollo Era,* op. cit., pp. xiii–xiv; on the concrete application of this principle in the planetary astronomy program see William E. Brunk, OHI by Tatarewicz, 21 July 1983 and 9 Aug. 1983, and the discussion of the Caltech proposal below.

22. Schorn, OHI; see also Harlan Smith, "Remarks Made at the Scientific Dedication of the 107-inch Reflector, October 30, 1969," in Carl Sagan, et al., eds., *Planetary Atmospheres,* IAU Symposium No. 40 (Dordrecht: Reidel, 1971), pp. 403–404; Harlan J. Smith, *Final Report NASr 242 NASA Texas 105-inch Telescope Project* (30 Dec. 1976), p. 2; David S. Evans and J. Derral Mulholland, *Big and Bright: A History of the McDonald Observatory* (Austin: University of Texas Press, 1986), pp. 146–52.

23. Schorn, OHI. On the opposition from within the lunar and planetary programs office, it is perhaps significant to note that the Planetary Atmospheres Subcommittee of the Space Sciences Steering Committee adopted a recommendation, "after considerable discussion," in February 1964, which stated: "This Subcommittee believes that the most urgently needed scientific data regarding Mars, as a prerequisite to direct investigation of the presence of life on the planet, are the mean density, composition and temperature near the surface. We further believe that there are no means of obtaining these data that are superior or even comparable in effectiveness to direct measurement with an entry probe. We, therefore, recommend increased emphasis and priority be placed on a program of development and testing of a feasible entry-probe system suitable for this purpose." The Subcommittee added that "indirect optical methods of estimating this surface pressure and atmospheric scale height from the Earth or from flybys are very unlikely to give a reliable value to better than a factor of two or three, the best value now lying close to the limiting value that would permit a high-drag device to

be used successfully." "Excerpts from the Minutes of the February 5, 6, and 7, 1964 Meeting of the Planetary Atmospheres Subcommittee," WNRC 255-74-663 : 50 "SL Chron."

24. Newell to files, "Conference Telephone Report with Professor Bacher of CIT, 23 July 1964," 27 July 1964, WNRC 255-79-0649: 14 (212) "Conference Reports by Newell. Period 1963–June 1967". Caltech opted to use its facility option for a space sciences building on campus. See also Brunk, OHI.

25. Newell to Seamans, 2 Oct. 1964.

26. Newell to Dubridge, 16 Nov. 1964, WNRC 255-79-0649: 2 (41) "SC". Essentially the same expressions of the NASA policy can be found in Naugle to Newell (drafted by Roman), 16 July 1964, "Proposed Air Force Lunar-Planetary Astronomical Observatory," op. cit.

27. Newell to Seamans, 2 Oct. 1964 255-67A601: 7 "Astronomy 1–2."

28. National Academy of Sciences, Committee on Astronomical Facilities, *Ground-Based Astronomy: A Ten Year Program* (Washington, D.C.: NAS, 1964), p. 26 and chart on p. 30.

29. Whitford Report, pp. 26, 9. The records of the Whitford Committee have not yet been organized, and were not available for consultation at the NAS Archives.

30. *Science* (13 Nov. 1964): 899–900.

31. Brunk, OHI, 9 Aug. 1983. The motivation for the figure of 25–50% guaranteed lunar and planetary time was the fact that due to planetary phenomena (risings, settings, apparitions) there are times of intense observation followed by long periods during which no planets are suitably placed for observation. Thus a facility dedicated entirely to lunar and planetary astronomy would lie idle during unfavorable times and would be inefficient. Yet during favorable apparitions NASA needed to be sure the facility would be available. Jay Bergstrahl, working with the McDonald 107-inch telescope in 1969, recalls periods during which Mars spectra were taken all night and Venus spectra taken all day; Bergstrahl, OHI 13 July 1983.

32. "Whitford Report Assessment," large folder in the NSF History Files, office of J. Merton England, NSF Historian, including "Draft Summary Minutes, Ad Hoc Interagency Group to Assess the Recommendations of the Whitford Report, First Meeting—February 24, 1965," ". . . Second Meeting—March 10, 1965," ". . . Third Meeting, March 17, 1965."

33. Haworth to Hornig, 5 Apr. 1965, transmitting "Assessment of the Recommendations of the Whitford Report, entitled, 'Ground-Based Astronomy—A Ten Year Program,' Report by an ad hoc Group representing those Federal Agencies which perform or support research in ground-based astronomy," including as appendices: Hornig to Haworth, 14 Jan. 1965 (requesting assessment); Dryden to Haworth, 23 Mar. 1965 (describing NASA policy, including copy of Naugle and Roman to Newell, 16 Mar. 1965, and 28 pp. of tables describing NASA present and anticipated support to astronomy).

34. Newell to Frank K. Edmondson, 20 Feb. 1981, FKE.

35. Haworth to Hornig, 15 Oct. 1965, transmitting National Science Board "approved suggestion" (27–28 May 1965) that NSF be the "principal Federal agent" for ground-based astronomy; see especially the several rough drafts, emendations, and concurrence sheets attached; Hornig to Haworth, 30 Nov. 1965, confirming the designation.

NASA had very good reason, especially at this time, to tread gingerly. In December 1965 the AAAS had released a damning report, "Integrity in Science," which took the overall space program severely to task for disrupting the orderly procedures and affairs of science. Critics of the space program, especially Apollo, were becoming increasingly vocal. The circumscription of NASA's involvement in astronomy was welcome to NASA as well, since the agency had no desire to become involved in ground-based astronomy any more than was required for its mission.

36. Hearth to Newell, et al., 6 Apr. 1964, "Program Recommendations." See also: Hearth to Newell, 3 Apr. 1964; "Report #1-Ad Hoc Planetary Study Group,"

3 Apr. 1964; Hearth to Ad Hoc Planetary Study Group, 2 June 1964, "Minutes of the May 26, 1964, Meeting . . ."; 10 Aug. 1964, "Minutes of August Meeting"; all in WNRC 255-67A601: 21 "SL Chron."

37. Hess to Newell, 12 Aug. 1964, transmitting "Future Goals of the Space Science Program," 11 Nov. 1964, WNRC 255-79-0649: 24 "Merker Reading File"; Hess to Webb, 23 Sept. 1964; Minutes, Special Meeting of the Space Science Board (with Webb), 11–12 Sept. 1964; Hess to Webb, 30 Oct. 1964, transmitting "National Goals in Space, 1971–1975," 28 Oct. 1964; all in NHOA, "Space Science Board." John Walsh, "Space: National Academy Panel Recommends Exploration of Mars as Major Goal in 1971–85 Period," *Science* 146 (20 Nov. 1964): 1025–27.

38. Brunk, OHI, 21 July 1983; folder, "Funding—Past," AR.

39. The 1969–73 oppositions, in addition to being optimum for ground-based astronomy, were also favorable for spaceflight missions: "lighting conditions, accessibility of biologically interesting landing areas, and velocity requirements are significantly better than they will be for another 12 to 14 years," Newell to Seamans, op. cit., p. 3. On the Soviet challenge, Newell wrote, "there is a chance of losing this important scientific opportunity for all time if we do not make the effort to capitalize on it early. We must expect that the Soviets will continue their pattern of three flights to Mars and Venus at each opportunity. They have already attem[p]ted ten planetary flights to our two," ibid.; see also Hearth to the record, 13 Oct. 1964, "Current Actions and Guidelines on the Post-1964 Mars Program"; Hearth to the record, 30 Sept. 1964, "Discussion of Planetary Program with the Associate Administrator, September 23, 1964"; Newell, "Scientific Requirements for Mars 1969 Orbiter and 1971 Lander Missions," 3-page attachment to Newell to Seamans, op. cit.

40. Brunk, OHI 9 Aug. 1983.

41. Grant NsG 451 (31 July 1963)—to support research in planetary astronomy and assist in the collection, interpretation, and dissemination of planetary information; Grant NsG(F) 15 (26 Oct. 1963)—Planetary Research Center construction; Brunk, "Trip Report"; Brunk to Nicks, 1 Mar. 1965, "Observations of Mars during the Current Opposition in March, 1965," AR.

42. Brunk, "Trip Report"; Newell to Naugle and Roman, 25 Nov. 1963, "Planetary Observatories [re: original Hynek-Tombaugh proposal]," WNRC 255-79-0649: 12 (183): "Chron File . . ."; Newell to Nicks and Liddel, 10 Dec. 1964, "Visit to New Mexico State College at University Park, New Mexico," WNRC 255-79-0649: 12 (185) "Chron File . . ."

43. Brunk to Nicks, 1 Mar. 1965, "Observations of Mars," op. cit.

44. Ezell and Ezell, *On Mars*, pp. 74–79; Colin S. Pittendrigh, Wolf Vishniac, and J. P. T. Pearman, eds., *Biology and the Exploration of Mars: Report of a Study Held under the Auspices of the Space Science Board . . . 1964–1965.* (Washington, D.C.: NAS Pub. 1296, 1966); R. N. Watts, Jr. "Mars Observations Wanted—Mariner 4 Probe," *Sky and Telescope* 29 (Mar. 1965): 150; W. J. Normylie, "Scientists Detail Mars Priority Argument; Voyager Program," *AWST* 82 (10 May 1965): 69; "On to the Red Planet: Mariner IV and Zond II," *Time* 84 (11 Dec. 1964): 76. See additional citations in Ezell, op. cit., note 65.

45. Ezell and Ezell, *On Mars*, p. 104; Naugle to "Mars Surface Pressure Determination Observers [A. Kliore, Von R. Eshleman, John S. Hall, David Rank, Edward C. Inn, Dennis C. Evans, Hyron Spinrad, Ron Schorn, Guido Munch, Tobias Owen, G. P. Kuiper, M. J. S. Belton, Reinhardt Beer]" 23 July 1965, in William E. Brunk, AR.

46. Ezell and Ezell, *On Mars*, p. 104 and sources in note 53, located in NHOA.

47. Ezell, *Mars*, and Koppes, *JPL*, op. cit.

48. Naugle to the record, 4 Oct. 1965; "Voyager Science Strategy," in box; "Voyager Documentation, 1962–67," NHOA.

49. Koppes, *JPL*, pp. 190–92; Ezell and Ezell, *On Mars*, p. 107.

50. National Academy of Sciences, Space Science Board, *Space Research: Di-*

rections for the Future, report of a study by the Space Science Board, Woods Hole, Mass., 1965. (Washington, D.C.: NAS pub. 1403, 1965), p. 9.

51. Ibid., p. 11.

52. Revised Minutes, seventeenth meeting, 12–13 Nov. 1965, NHOA.

53. Minutes, eighteenth meeting, 13–14 Feb. 1966, NAS:ADM:SSB:Meetings: 18th:Minutes:Draft.

54. John S. Hall, "Report on Conference on Martian Atmosphere and on Needs for Ground-based Planetary Astronomy," Attachment "A" to minutes of the twenty-first meeting, Space Science Board, 15–16 Dec. 1966. NAS:ADM:C&B:SSB:Meetings:21st:Minutes, p. 2.

55. Hall, "Report," pp. 4–5.

56. Hall, "Report," pp. 2, 6.

57. Brunk to Deputy Associate Administrator (Sciences), 17 Nov. 1966, "NASA Response to the Space Science Board's Recommendations for Further Studies of the Martian Atmosphere"; Newell to Hess, 28 Nov. 1966, transmitting "NASA Response"; Newell to Distribution, 9 Dec. 1966, transmitting "Recommendations for Further Studies of the Martian Atmosphere," 20 Oct. 1966 and list of participants; Newell to Hess, 13 Dec. 1966, with information on funding and telescopes supported by NASA; all of the preceding in William E. Brunk, AR. The major study which grew out of Hall's activities will be discussed in chapter 4, below.

58. Ray Newburn recalls a meeting at the NASA Western Operations Office in Santa Monica around the time of the Voyager troubles and the stalled Caltech negotiations. Caltech's Robert Leighton, JPL's Newburn and Robert Megrheblian, and Mount Wilson Director Ira S. Bowen met with Newell to try to work out the difficulties. "And everyone went right down the line in front of Homer Newell for the need for the planetary telescopes. And I remember Ira Bowen's telling him the story. He said 'you've got to have it. There is no other way you can get the time. We've turned down a proposal this year from the Astronomer Royal of Great Britain.' That made quite an impression on Homer Newell." Newburn, OHI, pp. 133–35.

59. Harlan J. Smith, *Final Report. NASr 242. NASA-Texas 105-Inch Telescope Project* (30 Dec. 1976).

60. Smith, "Final Report," pp. 5–7; Brunk to Nicks, 21 Feb. 1966, "Continuation of Funding of the Construction of NASA–McDonald 105-inch Telescope . . . ," AR.

61. The NASA policy on funding the instrument and not the facility led to an interesting quarrel with the NASA contracts officials over whether the dome was part of the instrument (since it opened and closed, rotated with the telescope, etc.) and should be funded by NASA or whether it was part of the building (since it was the roof) and hence should be the responsibility of the university.

62. Smith, "Final Report," pp. 7–9.

63. Smith, "Final Report," pp. 11–12.

64. Smith, "Final Report," pp. 95–97. as of the end of 1976, the final NASA cost for the instrument was $4.9 million, and the costs for supporting facilities totalled $4.9 million, of which $1.1 million was provided by NSF.

65. "Remarks Made at the Scientific Dedication of the 107-Inch Reflector, October 30, 1969," pp. 403–408 in Carl Sagan, et al., *Planetary Atmospheres;* Harlan J. Smith, "McDonald Observatory's 107-inch Reflector," *Sky and Telescope* 36/6 (Dec. 1968): 360–67; Robert G. Tull, "Planetary Spectroscopy with the 107-inch Telescope," *Sky and Telescope* 38/3 (Sept. 1969): 156–60.

66. Smith, "Final Report," pp. 13–14; Robert F. Allnutt to James G. Gehrig (drafted by Brunk), 18 Feb. 1970; Brunk to files, 16 Mar. 1970, "Corrective Actions to Minimize Damage to 107-inch Mirror at McDonald Observatory," AR.

67. Smith, "Final Report"; Brunk, ibid.

68. M. Mitchell Waldrop, "Mauna Kea (I): Halfway to Space," *Science* 214 (27 Nov. 1981): 1010–13, and "Mauna Kea (II): Coming of Age," *Science* 214 (4 Dec. 1981): 1110–14. Kevin Krisciunis, *Astronomical Centers of the World* (New York: Cambridge University Press, 1988), pp. 222–37, using documents from the Hawaii Chamber of Commerce, confirms a number of these points.

69. Newell to Thomas K. Smull (NASA Grants and Contracts), 8 Apr. 1964, "Proposals from the University of Hawaii," WNRC 255-79-0649: 12 (184) "Chron File . . ."

70. Waldrop, "Mauna Kea (I)," p. 1111; Mitsuo Akiyama, Executive Secretary Hawaii Island Chamber of Commerce to Newell, 17 Aug. 1964, enclosing copy of Kuiper's speech, WNRC 255-79-0649: 13 "R & A."

71. Urner Liddel to Robert Hiatt, Vice President for Academic Affairs, University of Hawaii, 20 Nov. 1964, WNRC 67A601: 7 "Astronomy."

72. To Kuiper, 10 Dec. 1964; Liddel to Kuiper, 15 Dec. 1964; Liddel to Menzel, 1 Feb. 1965; Liddel to Jeffries, 21 Jan. 1965: all of preceding in William E. Brunk, AR. Liddel to Menzel, 16 Dec. 1964, WNRC 67A601: 7 "Astronomy."

73. Newell to Hiatt, 16 Feb. 1965, WNRC 255-79-0649: 24 "Merker—Reading File."

74. Brunk, quoted in Waldrop, p. 1012; Waldrop, pp. 1011–12; Nicks to Newell, 31 Mar. 1965 (drafted by Brunk), AR.

75. Waldrop, p. 1012.

76. Waldrop, p. 1012; Nicks to Newell (drafted by Brunk), 16 Mar. 1966, "Continuation of Funding of Contract NSR 12-001-019 with the University of Hawaii for the Design and Construction of a 84-inch Telescope in the Hawaiian Islands," AR. Evenutally, the NASA 3-meter Infrared Telescope Facility was built on the summit, dedicated in July 1979, the first fully NASA-owned and operated telescope. Interestingly enough, this instrument too was justified in support of a Voyager project, but in this case the Voyager Jupiter-Saturn mission, related in name only to the Voyager Mars program.

77. Waldrop, "Mauna Kea (I)", pp. 1012–13; Brunk, "The Mauna Kea Observatory," 27 Aug. 1969, AR; Newell, "Speech . . . at Dedication of 88-Inch Telescope . . . June 26, 1970," WNRC 255-79-0649: 25 "AA Reading Files"; John T. Jeffries and William M. Sinton, "Progress at Mauna Kea Observatory," *Sky and Telescope* 36/3 (Sept. 1968): 140–45; "Mauna Kea Observatory Dedicated," *Sky and Telescope* 40/5 (Nov. 1970): 276–77. The final cost was about $3 million from NASA, $1 million from NSF, and $2 million from the University of Hawaii.

78. When Mars is at perihelion, and consequently of greatest angular size, it is nearly as far south of the ecliptic as it ever gets. This made the Hawaii site, at about 20-degrees latitude, particularly attractive. It is important to have the planets situtated as close to zenith as possible, especially for spectroscopy, so as to avoid observing them through a large air mass.

79. An excellent discussion of the various combinations and possibilities for planetary spectroscopy is Robert G. Tull, "Planetary Spectroscopy with the 107-inch Telescope," op. cit. A "flip-flop" aluminum-silver secondary mirror was postponed for the Texas telescope due to cost and time delays.

FOUR. THE PROGRAM MATURES (1965–1970)

1. Brunk to Newell, 18 May 1965, "Minor Construction Request 65-29 for 24-inch Observing Facility at Table Mountain," AR.

2. Brunk to Newell, 21 Apr. 1966, "Minor Construction Request 66-16 (Revised) for an Addition to the Existing 24-inch Observing Facility at the Table Mountain Site"; Newell to NASA Resident Office—JPL, 22 Apr. 1966 (approving all but coudé room); Brunk to NASA Resident Office—JPL, 1 Aug. 1966 (approving digging

out of coudé room); Brunk to Newell, 14 July 1967, (approving coudé spectrograph): all of preceding in Brunk, AR.

3. Newell to files, 30 June 1967, "Conference Report with W. H. Pickering, R. V. Meghreblian, R. H. Norton, Frank Goddard, R. J. Mackin of JPL with Myself re 'The Long Range Astronomy Interests of JPL,'" with attached MS, no date, R. L. Newburn and R. H. Norton, "The JPL Optical Astronomy Program," WNRC 255-79-0649: 14 (212) "Conference Reports by Newell . . ."

4. Brunk to Newell, 14 July 1967, "Proposed Astronomical Facilities on JPL/ Table Mountain as Reported in Your Memorandum to the Files Dated 30 June 1967," Ar.

5. See Newell, *Atmosphere,* pp. 259–60 and 267–69; correspondence and comments on the draft of the JPL chapter in WNRC 255-79-0649: 40 "Critics' Reviews"; Hall, *Lunar Impact,* pp. 240–55; Koppes, *JPL,* pp. 154–60, 207–11; H. L. Nieburg, *In the Name of Science* (Chicago: Quadrangle, 1966), pp. 234–44, and Oran Nicks's comments on Nieburg's assertions in reply to Newell to Nicks, 10 Nov. 1966, WNRC 255-79-0649: 24 "Merker—Reading File."

6. See, for example, Ronald A. Schorn, "The Spectroscopic Search for Water on Mars: A History," pp. 223–36 in Carl Sagan, et al., *Planetary Atmospheres;* G. E. Hunt and Jay T. Bergstralh, "Analysis of Spectroscopic Observations of Jupiter and the Variability of the Structure of the Visible Clouds," pp. 385–87 in A. Woszczyk and C. Iwaniszewski, eds., *Exploration of the Planetary System,* IAU Symposium no. 65 (Dordrecht: Reidel, 1974); "Table Mountain Observatory Backstops Mariners," NASA News Release 69-30, 23 Feb. 1969; Brunk to Nicks, 1 July 1969, "Evaluation of JPL Performance of Planetary Astronomy Tasks for the Period 1 January 1969 through 30 June 1969," 1 July 1969; Brunk, "JPL Performance Evaluation for the Period 1 July 1969 to 31 December 1969 Planetary Astronomy," 30 Jan. 1970; "JPL Performance Evaluation for the Period 1 January 1970 to 30 June 1970 Planetary Astronomy," 23 June 1970; all AR; Newburn, OHI, pp. 115–33.

7. Brunk to files, 13 Aug. 1965, "Meeting on 6 August 1965 with Dr. Horace Babcock and Mr. Bruce Rule of Mount Wilson and Palomar Observatories . . . ," AR.

8. Ira S. Bowen and Bruce H. Rule, "Palomar 60-inch Photometric Reflector," *Sky and Telescope* XXXII (Oct. 1966); unpaginated offprint in Brunk, AR; Brunk to Associate Deputy Administrator, 8 Oct. 1970, "Background Material for Remarks at Dedication of the California Institute of Technology 60-inch Telescope: NASA Involvement in Ground-based Astronomy," AR; Robert B. Leighton, et al., "Proposals and Progress Reports Concerning NASA Grant NsG 426," large binder in Gerry Neugebauer, SAOHPWF.

9. Lowell Observatory Receives NASA Grant," NASA News Release 63-235, 21 Oct. 1963; Brunk to Hall, 20 Apr. 1965, AR; Brunk to Dollfus, 2 July 1965, AR.

10. Brunk to Liddel, 28 Oct. 1965, "Continuation of NASA Grant NsG 451 . . ."; Brunk to Nicks, 30 Nov. 1966, "Continuation of Funding . . . ," AR.

11. John S. Hall, "Report on Conference on Martian Atmosphere and on Needs for Ground-based Planetary Astronomy," attachment "A" to minutes of the twenty-first meeting, Space Science Board, 15–16 Dec. 1966, NAS:ADM:C&B:SSB:Meetings:21st:Minutes, p. 5.

12. Ibid. Don P. Hearth to John Naugle, 21 Mar. 1968, "Funding of Lowell Observatory to conduct a Synoptic Photographic Planetary Patrol . . . ," drafted by Brunk, AR (at this time Newell had become Associate Administrator of NASA, Naugle had moved into Newell's old job, and Hearth had become director of Lunar and Planetary Programs).

13. "Lunar and Planetary Notes from Prague," *Sky and Telescope* (Jan. 1968): 7–8 (Report on the Aug. 1970 meeting of Commission 16 of the IAU); J. L. Inge,

et al., "A New Map of Mars from Planetary Patrol Photographs," *Sky and Telescope* (June 1971):336–39; "International Planetary Patrol Results," *Sky and Telescope* (Nov. 1971): 261–62; W. A. Baum, et al., "The International Planetary Patrol Program, *Icarus* 12 (May 1970): 435–39.

14. W. A. Baum and L. J. Martin, "Cloud Motions on Mars," pp. 320-28 in Carl Sagan, et al., *Planetary Atmospheres;* W. A. Baum, et al., *Mars Cloud Survey Report no. 1* (Flagstaff, Ariz.: Lowell Observatory Planetary Research Center, 1967); L. J. Martin and W. A. Baum, *A Study of Cloud Motions on Mars, Final Report, Part B* (Flagstaff, Ariz.: Lowell Observatory Planetary Research Center, 1969); W. A. Baum, "Results of Current Mars Studies at the IAU Planetary Research Center," pp. 241–53 in *Exploration of the Planetary System,* op. cit.; for the interaction of the planetary patrol results with landing site selection and other aspects of the Viking Mars Landing mission see Ezell and Ezell, *On Mars,* chap. 10.

15. Harry Hess, "Foreword," in National Academy of Sciences, Space Science Board, Panel on Planetary Astronomy, *Planetary Astronomy: An Appraisal of Ground-Based Opportunities* (Washington, D.C.: NAS, 1968). Other members of the panel: J. W. Chamberlain, W. C. DeMarcus, R. Hide, G. P. Kuiper, B. Mason, C. H. Mayer, B. C. Murray, W. A. Noyes, J. Oro, T. C. Owen, G. H. Pettengill, D. H. Rank, E. Roemer, R. L. Wildey, N. W. Hinners (consultant), W. E. Brunk and U. Liddel (contributors), B. N. Gregory (study director). The records of the Hall panel have not yet been organized, and are not available at the NAS Archives.

16. Hall Report, pp. 1–6.

17. Hall Report, pp. 7–8.

18. Hall Report, pp. 18–25.

19. Hall Report, pp. 26–38.

20. Hall Report, p. 50; Hall, "Report on Conference on Martian Atmosphere and on Needs for Ground-based Planetary Astronomy," attachment "A" to minutes of the twenty-first meeting, Space Science Board, 15–16 Dec. 1966, NAS:ADM: C&B:SSB:Meetings:21st:Minutes, p. 2.

21. Hall Report, pp. 64–68. Additional results of the survey will be discussed in detail and compared with other similar evidence in the following chapter.

22. Brunk, "Proposed SR&T Seed Grants, Planetary Astronomy," 27 Oct. 1967, AR.

23. Brunk, "Planetary Astronomy—FY 1971 New Starts," 3 Sept. 1969, and a seven-page narrative justification; Brunk, "Material requested for the record . . . ," 6 Apr. 1971; AR.

24. Ezell and Ezell, *On Mars,* pp. 110–19; Luther J. Carter, "Planetary Exploration: How to Get by the Budget-Cutters?" *Science* 158 (24 Nov. 1967): 1025–28.

25. Ezell and Ezell, *On Mars,* pp. 121–22; Luther J. Carter, "Space Budget: Congress Is in a Critical, Cutting Mood," *Science* 157 (14 July 1967): 170–73; Carter, "Space: 1971 Mariner Mission Knifed by Budget-Cutters," *Science* 157 (11 Aug. 1967): 658–60.

26. Carter, "How to Get by the Budget-Cutters?"

27. President's Science Advisory Committee, Joint Space Panels, *The Space Program in the Post-Apollo Period* (Washington, D.C.: The White House, Feb. 1967), pp. 17, 21.

28. Ibid., pp. 17–18; the competition between Newell's and Mueller's offices for funds and support will be discussed below. It had long disturbed many scientists, and the difficulties came to a head in 1969 with post-Apollo decisions nearing resolution. See below.

29. Ibid., pp. 18–19. The apparent lack of coordination, and even competition, between the Space Science Office and the Office of Manned Space Flight at NASA was a continuing source of confusion and dismay to the various scientific advisory bodies to NASA.

30. "Space Science Board Statement on Post-Apollo Goals (1968-1975)," 29

July 1966; Hess to Webb, 29 July 1966; NAS:ADM:C&B:SSB:Policy:Post Apollo Goals 1968–1975:Statement: Transmittal to NASA.

31. Space Science Board, Ad Hoc Committee on Small Planetary Probes, "Exploration of Space with Sub-Voyager Systems," Memorandum Report, 1967. See also Van Allen quoted in Carter, "Space: 1971 Mariner Mission Knifed by Budget-Cutters," p. 659.

32. The planetary astronomy support was officially given program status in 1969.

33. Ezell and Ezell, *On Mars,* pp. 131–36; Carter, "Budget Cutters," p. 1025; Webb's proposals initially had the opposite effect on NASA's Lunar and Planetary Missions Board. The members were most upset at being only casually consulted at the last minute before Webb revealed the proposals in response to Anderson's question. They were equally upset at the emphasis on Mars and the persistence of the Saturn V–class Voyager-type mission, since they feared it would grow into another sprawling program for which other missions would eventually have to be sacrificed.

34. Space Science Board, *Planetary Exploration, 1968–1975,* report of a study by the Space Science Board (Washington, D.C.: NAS, 1968).

35. Ezell and Ezell, *On Mars,* p. 144.

36. Space Science Board, "Planetary Exploration," pp. 3, 16. NASA officials did not appreciate the SSB telling them how to allocate funds, a function they thought should be reserved only for the agency itself. The mention of funds for manned planetary exploration refers to planning in the Office of Manned Spaceflight at this time to attempt a manned Mars flyby using modified Apollo hardware. See below. See also Ezell, "Man on Mars."

37. SSB, "Planetary Exploration," pp. 14, 23–25.

38. SSB, "Planetary Exploration," pp. 12–13. In this case the Board was clearly out of line with NASA's budget. While $30 million was a drop in the Apollo bucket, the entire planetary astronomy program budget was around $3.9 million for 1968. Nonetheless, Brunk and the NSF later worked out arrangements to resurface the Arecibo antenna and provide it with a high-powered transmitter for planetary radar for a fraction of the cost of a new facility.

39. SSB, "Planetary Exploration," pp. 14–15. The use of earth satellites and customizing an astronomy satellite for planetary observations proved to be more intractable. The astronomy satellites wre designed for studies of point-surfaces which remained fixed in right ascension and declination. Observations of planets required tracking in those coordinates to correct for the planet's motion in its orbit as well as detectable parallax as the satellite moved in its own orbit. The size of the planetary disc and guide star tracking also presented serious difficulties. These difficulties in the use of astronomy satellites for planetary research, and in accommodating both planetary and stellar/galactic requirements in the same satellite observatory, will be the subject of a forthcoming investigation.

40. Norriss S. Hetherington, "Winning the Initiative: NASA and the U.S. Space Science Program," *Prologue* (Summer 1975): 99–107; Hetherington, "The Evolution of NASA's Science Program, 1958–1960," NASA Historical Note HHN-123, August 1972, NHOA; Charles M. Atkins, "NASA and the Space Science Board of the National Academy of Sciences," NASA Historical Note HHN-62, September 1966, NHOA.

41. Barry Rutizer, "The Lunar and Planetary Missions Board," NASA Historical Note HHN-138, 30 Aug. 1976, NHO, pp. 16–17.

42. Rutizer, pp. 22–24; Ezell, *Mars,* pp. 208–215; minutes of LPMB Meetings, NHOA.

43. Robert O. Doyle, ed., *A Long-Range Program in Space Astronomy,* position paper of the Astronomy Missions Board (Washington, D.C.: NASA SP-213, July 1969 [published November]), pp. iii, 13.

44. Ibid., pp. 14–15, "A Program for Planetary Astronomy from Space Tele-

scopes," report of the Planetary Astronomy Panel, 24 Dec. 1968, revised 15 Apr. 1969, pp. 177–201. Panel members were: J. W. Chamberlain (Chmn.), D. C. Evans, D. M. Hunten, F. J. Low, G. Munch, G. C. Pimentel, Harlan J. Smith, and W. E. Brunk.

45. Ibid., "Ground-Based Astronomy in an Integrated National Program," report of the Ground-Based Astronomy Working Group, Sept. 1968, revised 1 May 1969 and 4 June 1969. Members were: C. R. C. O'Dell (Chmn.), B. F. Burke, A. D. Code, W. Furor, W. A. Fowler, A. E. Whitford.

46. Ibid., p. 257.

47. Ibid., pp. 258–59.

48. Ibid., pp. 265–67.

49. Newell, *Atmosphere,* pp. 286–89, 219; Edward C. Ezell, "Man on Mars: The Mission That NASA Did Not Fly," *STTH: Science, Technology, and the Humanities* 3 (1980): 21–34; John M. Logsdon, *Space Stations: A Policy History,* prepared for Johnson Space Center under contract NAs9-16461, no date, but 1982?

50. Robert W. Smith, *The Space Telescope: A Study of NASA, Science, Technology, and Politics* (Cambridge: Cambridge University Press, 1989).

51. *Space Task Force—Report* (no date, but Jan. 1969) cited in Logsdon, *Space Stations: A Policy History;* see also Russell Drew, "The White House and Space Policy," lecture delivered at the Seminar on Space Policy, National Air and Space Museum, 8 Apr. 1982, pp. 32–39, transcript in SAOHPWF.

52. Newell, *Atmosphere,* p. 288; John M. Logsdon, "The Space Program during the 1970s: An Analysis of Policymaking" (Washington, D.C.: Comment Draft, 1974), pp. 3–7. In the narrative I generally follow Logsdon, supplemented by other sources.

53. Logsdon, "The Space Program," pp. 7–32; *America's Next Decade in Space: A Report for the Space Task Group* (Washington, D.C.: NASA, 1969); George P. Mueller, "An Integrated Space Program for the Next Generation," *Astronautics and Aeronautics* (Jan. 1970) and Mueller, "Post-Apollo Revisited," *Astronautics and Aeronautics* (Jul./Aug. 1979): 24–31. See also Russell Drew, "The White House and Space Policy," op. cit., pp. 44–47.

54. Logsdon, "The Space Program," pp. 33–48; *The Post-Apollo Space Program: Directions for the Future,* Space Task Group report to the President (Sept. 1969).

55. Newell, *Atmosphere,* pp. 218–19, 404–405; Rutizer, *LPMB,* pp. 27–29.

56. Newell to Special Assistant to the Associate Administrator, 6 Nov. 1969, "Assemble a File," WNRC 255-79-0649: 24 "AA Chron Files."

57. Newell to Fletcher, no date, but 3 Dec. 1971, "Relations with the Scientific Community and the Space Science Board," WNRC 255-79-0649: 25.

58. Newell, *Notes on Science in NASA,* 20 pp. plus attachments, 14 Nov. 1969, WNRC 255-79-0649: 24 "AA Chron Files," p. 1. The unhappiness had been growing for some time. See Newell to Paine, 16 May 1969, "Notes for Meeting with NASA Advisors, Sunday, May 18, 1969, at KSC," WNRC location as above, in which Newell briefs Paine on the chief concerns of the various advisory bodies. Newell continued to believe that the problems with the LPMB were more a matter of perspective than of the specific recommendations in the STG report; Newell, *Atmosphere,* pp. 218–19, 404–405.

59. Newell, *Notes on Science in NASA,* pp. 3, 13; see also Newell, "Spadework for Exploration of the Solar System," pp. 47–56 in George W. Morganthaler and Robert G. Mora, eds., *Unmanned Exploration of the Solar System* (North Hollywood, Calif.: Periodicals, Inc., American Astronautical Society, 1965).

60. Newell, *Notes on Science in NASA,* p. 17.

61. Newell to Paine, 7 Nov. 1969, "NSF/NASA Support to Astronomy," WNRC 255-79-0649: 24 "AA Chron Files." NSF had recently had two facilities previously supported by the Department of Defense, the Arecibo radio telescope in Puerto Rico and the MIT Haystack Radar facility "left on its doorstep."

62. Henry J. Smith to the files (drafted by Brunk), 11 Dec. 1969, "National

Science Foundation (NSF)–National Aeronautics and Space Adminstration (NASA) Coordination Meeting on the Support of Ground-Based Astronomy," AR.

63. Paine to McElroy, 12 Dec. 1969, copy in NSF History Files.

64. See Newell, *Atmosphere,* pp. 288–91.

65. Logsdon, "The Space Program," pp. 48–69; Ezell and Ezell, *On Mars,* pp. 274–78; John Naugle, "Planning Planetary Missions," lecture delivered at the Seminar on Space Policy, National Air and Space Museum, 4 Mar. 1982, transcript in SAOHPWF.

66 Logsdon, "The Space Program," pp. 70–71; Newell, *Atmosphere,* pp. 288–89. Nixon's full statement of 7 Mar. 1970 is reproduced in Newell's book as appendix J.

67. Newell, *Atmosphere,* p. 288.

68. Newell to the record, 30 Apr. 1970, WNRC 255-79-0649: 25 "AA Reading Files."

69. Newell, *Atmosphere,* pp. 289–90.

70. R. W. Nichols, "Mission Oriented R & D," *Science* 172 (2 Apr. 1971): 29–33; J. Maddox, "American Science: The Endless Search for Objectives," *Minerva* (1972): 129–40.

71. Brunk to files, 15 Dec. 1969, "Notes of Telephone Conversations, Henry Smith and Robert Fleischer, National Science Foundation (NSF), 15 December 1969"; Brunk to files, 18 Dec. 1969, "Telephone Conversation on December 17, 1969 with Paul Sebring, Manager, Haystack Facility, Lincoln Laboratory, M.I.T."; Brunk to distribution, 10 June 1970, "Funding of Radar Astronomy at the MIT Haystack Antenna for the Fiscal Year 1971"; all in Brunk, AR. Packet of correspondence, routing slips, and memos attached to Newell to Jerome B. Wiesner, 31 Mar. 1972 and Newell to (NASA Administrator) Fletcher, 8 June 1972 with attachments, all in WNRC 255-79-0649: 26 "AA Reading File."

72. Brunk to Acting Director of Science, Office of Space Science, 16 Dec. 1965, "Meeting of the Arecibo Evaluation Panel at the Dorado Beach Hotel, Puerto Rico, 30 November–1 December 1965"; Brunk to Naugle, 24 Feb. 1970, "Congressional Backup Statement for Planetary Astronomy"; Brunk to NSF, 7 Sept. 1971, "Authorization to Proceed with Phase I Activities under NASA Purchase Order No. W-13, 359"; "Memorandum of Agreement between NASA and NSF for the Addition of a High-Power S-Band Radar Capability and Associated Additions and Modifications to the Suspended Antenna Structure of the National Astronomy and Ionosphere Center at Arecibo," 24 June 1971; Brunk to Associate Administrator for Space Science, 12 July 1974, "Arecibo Observatory Upgrading"; all in Brunk, AR.

73. M. Mitchell Waldrop, "Mauna Kea (II): Coming of Age," *Science* 214 (4 Dec. 1981): 1110–14 and numerous papers in the chronological files and subject files of Brunk, AR.

74. See below, chapter 5. For the truly grandiose ambitions of Paine, he would have had to court and keep the favor of the executive branch as well as Congress for a long stretch of time. The only time this had been done in the space context was during the Apollo program, and even then the alliance did not last.

75. See Newell, *Atmosphere,* pp. 288–91, and Paine's handwritten comments on Newell's draft manuscript in WNRC 255-79-0640: 40 "Critics' Reviews." See also John Naugle, "Planning Planetary Missions," lecture at the National Air and Space Museum Seminar on Space Policy, 4 Mar. 1982, pp. 21, 50–51, and Logsdon, "The Space Program."

FIVE. CONCLUSION: A HOME FOR
PLANETARY SCIENCE?

1. Eli Ginzberg, et al., *Economic Impact of Large Public Programs: The NASA Experience* (Salt Lake City: Olympus, 1976), "Transformations of a Science: NASA's

Impact on Astronomy," pp. 82–83. This study made use of National Academy of Sciences reports and undocumented interviews. Judging from the phrasing in the report and the dates of the interviews, I believe it highly likely that Homer Newell and others at NASA were the source for some of these assertions.

The resources—money and facilities, including spacecraft missions—made available by the government to planetary research were immense in terms of the field, but relatively meager compared with other areas of physics. Nonetheless, the same patterns of activity and dilemmas for the practitioners arose. See Paul Forman, "Behind Quantum Electronics: National Security as Basis for Physical Research in the United States, 1940–1950," *Historical Studies in the Physical and Biological Sciences* 18/1 (1987): 149–229, as well as the other articles in the same volume.

2. For the history of X-ray astronomy see Richard F. Hirsh, *Glimpsing an Invisible Universe: The Emergence of X-Ray Astronomy* (Cambridge: Cambridge University Press, 1983, in press).

On the extensive literature concerning specialty studies in the sociology of science see Daryl E. Chubin, "The Conceptualization of Scientific Specialties," *Sociological Quarterly* 17 (Autumn 1976): 448–76.

3. Ginzberg, "NASA's Impact on Astronomy," p. 92.

4. Newell, *Atmosphere*, p. 408.

5. Ibid., p. 328. But see especially my discussion of Newell's activities with the American Geophysical Union below. Newell used his AGU papers and notes in writing *Beyond the Atmosphere*.

6 "The wide range of problems served to draw together workers from a number of disciplines. Astronomers found themselves working with geophysicists who came to dominate the field of planetary studies that had once been the sole purview of the astronomers," ibid., p. 329.

7. Much of this is only oral history testimony, supplemented by a comparison of overall NSF funding in various fields with that in astronomy. Some correspondence among astronomers associated with the development of the first national observatory and the beginnings of the NSF astronomy program in the mid-1950s (Kitt Peak) bears this out. The Whitford Report, discussed in chapter 3 above, attests to how ill-prepared astronomers were for the sudden demand on facilities and people. See Robert W. Smith, *The Space Telescope*.

8. Solar System Exploration Committee, NASA Advisory Council, *Planetary Exploration through the Year 2000: Part One: A Core Program* (Washington, D.C.:GPO, 1983), p. 50.

9. Working Group on Planetary Science, Astronomy Survey Committee, National Academy of Sciences, *Challenges to Astronomy and Astrophysics: Working Documents of the Astronomy Survey Committee* (Washington, D.C.: NAS, 1983), pp. 98–99.

10. National Academy of Sciences, Space Science Board, Panel on Planetary Astronomy, *Planetary Astronomy: An Appraisal of Ground-Based Opportunities* (Washington, D.C.: NAS, 1968). See also chapter 5 of this study.

11. Jastrow and MacDonald to Berkner, 10 Dec. 1959; Berkner to Newell, 18 Dec. 1959; Newell to Berkner, 27 Dec. 1959; all in WNRC 255-79-0649: 4 (66) "AGU Planning Committee on Planetary Science: 1960" (two folders are labeled 66). Unless otherwise noted, all primary materials cited will be in this location. Other members of the PCPS were Philip Abelson, Leroy Alldredge, Thomas Gold, G. J. F. MacDonald, Hugh Odishaw, Allan Shapley, Harry Wexler, and Charles A. Whitten.

12. Sections of the AGU were: Geodesy; Seismology; Meteorology; Geomagnetism and Aeronomy; Oceanography; Volcanology, Geochemistry, and Petrology; Hydrology; and Tectonophysics.

13. H. E. Newell, R. Jastrow, G. MacDonald, "A Home for Planetary Sciences," prepared for the President's Page in *Transactions of the American Geophysical*

Union, draft MS (June 1960) in above location. Published in *Transactions of the American Geophysical Union* 41 (Sept. 1960): 407–409.

14. Newell, et al., "A Home for Planetary Sciences."

15. Newell to PCPS members, 5 Nov. 1960; "Summary of Comments from Members of AGU Planning Committee Concerning the Advantages and Disadvantages of a Planetary Science Section within the AGU (Ans. to 5 Nov. ltr.)"; Newell to PCPS members, 20 Feb. 1961 and "Draft Report to the American Geophysical Union Executive Committee on the Status of Planetary Sciences Within the AGU," 20 Mar. 1961; minutes of the meeting of the Executive Committee of the Council, American Geophysical Union, 25 Mar. 1961, pp. 12–13 and attachment 6c (in folder 61, "American Geophysical Union (AGU) General. 1960–1961").

16. Attachment 6c, report of the PCPS.

17. Minutes of the Executive Committee of the Council of the AGU, 11 Nov. 1961, p. 9.

18. Abelson to distribution, 22 May 1961; Dessler to Abelson, no date, but ca. 29 May 1961; Nieburger to Abelson, 31 May 1961; all in folder 61.

19. "Committee to Study Planetary Science Proposal, Philip H. Abelson, Chairman," attachment to minutes of the Executive Committee of the Council, AGU, 11 Nov. 1961.

20. Newell to Thomas F. Malone (succeeded Berkner as AGU president), 29 Mar. 1962; Newell to Malone, 20 Apr. 1962, with attached "Report of Activities of the Planning Committee on Planetary Science, 20 April 1962"; minutes of the Council meeting, 25 Apr. 1962.

21. Malone to Newell, 30 Nov. 1962; Newell to Jastrow, 27 Nov. 1962, with attached "Conference Report, Conference on Organization and Planning for the New Section on Planetary Sciences . . . [Newell and Jastrow], 13 November 1962."

22. To some extent this reorganization represented a return to the more traditional structure of the AGU, and is seen by some participants in the debate as an indication that Newell et al. were wrong in the first place. See Alex Dessler, OHI by Tatarewicz, 2–3 Apr. 1984, SAOHP.

23. "By-Laws of the Division of Planetary Sciences, American Astronomical Society, 15 May 1969," Article 3; Chamberlain to distribution, 10 Dec. 1968; "Abstracts of Papers Presented at the Pre-Inaugural Meeting of the Planetary Division of the AAS," *Bulletin of the American Astronomical Society* 1/2 (1969): 213–19. Copies of certain DPS materials were kindly provided by Dale P. Cruikshank, Secretary-Treasurer, and David Morrison, former Secretary-Treasurer.

The surviving archives of the division are being assembled.

24. Frank Drake, Chairman, "Proposal for Change in By-Laws of the Division of Planetary Sciences"; Peter M. Millman, "Meeting Review: The Fourth Annual Meeting of the AAS Division for Planetary Sciences, Tucson, Arizona, March 20–23, 1973," *Icarus* 20 (1973): 346–55.

To some extent these actions by the parent society represented its response to pressure from other space-related areas of astronomy—solar physics, high-energy, and dynamical astronomy (celestial mechanics).

25. Report of the Publications Subcommittee, ca. March 1973. Members were Clark R. Chapman, David Morrison, R. Reynolds.

26. Report of the Publications Subcommittee; Millman, "Meeting Review," op. cit.. The six journals were: *Icarus, Astrophysical Journal, Science, Journal of Geophysical Research, Astronomical Journal,* and *Journal of Atmospheric Sciences.*

27. David O. Edge and Michael J. Mulkay, *Astronomy Transformed: The Emergence of Radio Astronomy in Britain* (New York: Wiley, 1976), ch. 10.

28. For similar activities at the same time involving solar astronomy see Karl Hufbauer, *Exploring the Sun,* forthcoming. The debate within the American Astronomical Society over establishing any section or division structure at all was, by some accounts, lengthy and difficult.

29. See Bruno Latour, *Science in Action: How to Follow Scientists and Engineers through Society* (Cambridge, Mass.: Harvard University Press, 1977). Thomas P. Hughes, "The Seamless Web: Technology, Science, Etcetera, Etcetera," *Social Studies of Science* 16 (1986): 281–92: "the system builders were no respecters of knowledge categories or professional boundaries" (p. 285). Scientists, like historians or sociologists, might make a distinction between performing the substance of their craft and "administrative" activities. But a striking and consistent feature of correspondence between scientists and scholars, as revealed in archival collections, is how easily they mix discussion of the two "separate" activities.

30. I am indebted to Thomas F. Gieryn for many cautionary discussions on the validity and usefulness of survey and demographic data and analyses of problem choice. See his "Space Sociology? Comment on Hufbauer and Tatarewicz," in Martin J. Collins and Sylvia D. Fries, eds., *Proceedings of the Joint National Air and Space Museum–NASA History Office Conference on Space History,* June 1986 (forthcoming).

31. Gieryn assembled a machine-readable data base consisting of biographical information and publication histories for 2,308 American astronomers and astrophysicists professionally active between 1950 and 1975. For more information on the terminology, the data base, and the procedure used to construct the astronomical specialities see Thomas F. Gieryn, "Patterns in the Selection of Problems for Scientific Research: American Astronomers, 1950–75" (Ph.D. diss., Columbia University, 1979); for an example of how these data have been used see Gieryn, "problem Retention and Problem Change in Science," *Sociological Inquiry* 48 (1978): 96–115.

For additional information on how the data were manipulated to produce the graphs discussed here see Joseph N. Tatarewicz, "'Where are the people who know what they are doing?' Space Technology and Planetary Astronomy, 1958–1975" (Ph.D. diss., Indiana University, 1984); Tatarewicz and Gieryn, "Federal Funding and Planetary Astronomy, 1950–1975," in the National Science Foundation Division of Policy Research and Analysis, *Proceedings of the Workshop on Federal Funding and Knowledge Growth in Subfields and Specialties of Science, May 1983;* and Tatarewicz, "Federal Funding and Planetary Astronomy," *Social Studies of Science* 16/1 (Feb. 1986): 80–103.

32. For similar analyses which make use of absolute numbers of publications see the report of the Panel on Organization, Education, and Personnel of the [1980] Astronomy Survey Committee, *Astronomy and Astrophysics for the 1980s, vol. 2, Reports of the Panels* (Washington, D.C.: NAS, 1983), pp. 334–438; this report contains a wealth of demographic and other data plus a detailed bibliography. The report of the Working Group on Planetary Science, *Challenges to Astronomy and Astrophysics for the 1980s: Working Documents of the Astronomy Survey Committee* (Washington, D.C.: NAS, 1983), also contains some bibliometric data, as well as descriptions of the planetary science community and its development during the 1970s. Certain of the findings presented here are corroborated in the above studies.

33. Derek deSolla Price, *Little Science, Big Science: Science Since Babylon.*

34. National Academy of Sciences, Astronomy Survey Committee, *Astronomy and Astrophysics for the 1970s* (Washington, D.C.: NAS, 1972), vol. 2, p. 316, table 9.14.

35. National Academy of Sciences, Astronomy Survey Committee, *Astronomy and Astrophysics for the 1970s,* vol. 1, p. 60, table 3.

36. National Academy of Sciences, Space Science Board, Panel on Planetary Astronomy, *Planetary Astronomy: An Appraisal of Ground-Based Opportunities* (Washington, D.C.: NAS, 1968), p. 62, table 6.

37. National Academy of Sciences, Astronomy Survey Committee, *Astronomy and Astrophysics for the 1980s* (Washington, D.C.: NAS, 1983), vol. 2, p. 423, table 6.A.3.

38. Hall Report, p. 66.

39. Hall Report, p. 67.

40. "Students holding traineeships were found to be almost completely isolated from NASA. They were intellectually aware that their support came from the space program but had developed no ties or identification with NASA. One result of this was found in the Scott Study." *A Study of NASA University Programs* (Washington, D.C.: NASA SP-185, 1968), p. 51. The Scott Study mentioned is William E. Scott, "Federal Support to Graduate Students: A Comparison of NASA Traineeships With Research Assistantships" (M.S. thesis, Sloan School of Management, MIT, 1968).

41. See also "Levers of Control," below.

42. Tatarewicz, "Space Technology," pp. 380–81; Hirsh, op. cit.

43. Tatarewicz, "Space Technology," pp. 381–82.

44. Tatarewicz, "Space Technology," pp. 382–83.

45. Tatarewicz, "Space Technology," pp. 383–84. Due to the particulars of the data available from NASA on funding of individual principal investigators these are surely low estimates.

46. In the 1970–75 period there were around 150 to 200 scientists publishing exclusively on planetary topics. The proportion of these using ground-based techniques is hard to judge, but probably represents a significantly smaller number. The 1980 Working Group on Planetary Science estimated that of the 15 to 20 planetary science Ph.D.s produced each year, less than five were planetary *astronomers*. If that proportion is representative, then the number of planetary astronomers in the 1970–75 period can be estimated at around one hundred or less.

47. I was unable to find comparable information on geophysicists and earth scientists, which would have made an intersting comparison.

48. U.S. Congress, Office of Technology Assessment, *Space Science Research in the United States*, p. 26.

49. Ibid.

50. U.S. Congress, Office of Technology Assessment, *Space Science Research in the United States: A Technical Memorandum* (Washington, D.C.: GPO, Sept. 1982), pp. 5–8; Mitchell Waldrop, "Planetary Science on the Brink Again" *Science* 206/4424 (14 Dec. 1979): 1288–89; R. Jeffrey Smith, "Uncertainties Mark Space Program of the 1980s," ibid., 1284–86; "The NASA Budget: Planetary Panic," *Science News* 120 (26 Oct. 1981): 260. See also the Introduction, above.

51. Hall Report, pp. 67–68.

52. Thomas F. Gieryn, "Problem Retention and Problem Change in Science," *Sociological Inquiry* 48 (1978): 96–115.

53. Richard F. Hirsh, "Science, Technology, and Public Policy: The Case of X-ray Astronomy, 1959–1972" (Ph.D. diss., University of Wisconsin, 1979), p. 303; see also Hirsh, *Glimpsing an Invisible Universe*.

54. See Brunk to M. J. S. Belton, 18 Nov. 1975, AR, concerning problems at one of the NASA-sponsored telescopes: "The combination of a small planetary staff and pressures for more observing time from the larger number of non-planetary staff members are the basic causes of the problem. . . . The present university-supported staff has predominantly stellar interests."

55. Wolfgang van den Daele and Peter Weingart, "Resistance and Receptivity of Science to External Direction: The Emergence of New Disciplines under the Impact of Science Policy," in Gerard Lemaine, et al., eds., *Perspective on the Emergence of Scientific Disciplines* (The Hague: Mouton, 1976), p. 247. The rigid internal-external distinction between science and the environment in which it operates has recently come under severe questioning in the sociology of science. However, for the present purposes NASA was definitely an external force as viewed by the astronomical community, which controlled access to the ground-based instruments. For challenges to this internal-external distinction see especially Latour,

Science in Action, and Michel Callon, "Society in the Making: The Study of Technology as a Tool for Sociological Analysis," pp. 83–103 in Wiebe E. Bijker, Thomas P. Hughes, and Trevor Pinch, eds., *The Social Construction of Technological Systems: New Directions in the Sociology and History of Technology* (Cambridge, Mass.: MIT Press, 1987).

56. Ibid., p. 269.

57. For discussion of NASA's University Program and the frustrated aspirations of Administrator James Webb see W. Henry Lambright, *Governing Science and Technology* (New York: Oxford University Press, 1976) and *Presidential Management of Science and Technology: The Johnson Presidency* (Austin: University of Texas Press, 1985). See also Homer Newell, *Beyond the Atmosphere* and James Webb's preface to Arnold Levine, *Managing NASA in the Apollo Era*.

58. Brunk to Belton, 18 Nov. 1975, op. cit.

59. Michael J. S. Belton, "Planetary Astronomy with the S[pace] T[elescope]," in *Scientific Research with the Space Telescope* (Washington, D.C.: NASA CP-2111, 1979).

60. Working Group on Planetary Science, Astronomy Survey Committee, National Academy of Sciences, *Challenges to Astronomy and Astrophysics: Working Documents of the Astronomy Survey Committee* (Washington, D.C.: NAS, 1983), p. 108.

61. Van den Daele and Weingart, op. cit., p. 273.

62. Richard Berendzen, "On the Career Development and Education of Astronomers in the United States" (Ph.D. diss., Harvard University, 1968).

63. G. Nigel Gilbert, "The Development of Science and Scientific Knowledge: The Case of Radar Meteor Research," op. cit.

64. Edge and Mulkay, *Astronomy Transformed*, p. 22.

65. Ronald E. Doel, "Unpacking a Myth."

66. Edge, "The Sociology of Innovation," p. 335.

67. Edge and Mulkay, *Astronomy Transformed*, p. 359.

68. Through the years the planetary scientists took greater pains to show how their specialty was relevant to the overall goal of understanding stellar evolution and cosmology. In a recent report of the Solar System Exploration Committee, the authors note: "If we are to understand star formation fully, we must know how solar systems form. . . . The evolution of the solar nebula from a cloud of matter in the interstellar medium to the solar system we find today cannot be traced without understanding the most massive planets, accounting for their differences from each other as well as from their tiny cousins in the inner solar system." NASA Advisory Council, Solar System Exploration Committee, *Planetary Exploration Through the Year 2000, Part One, A Core Program* (Washington, D.C.: GPO, 1983): 129.

69. Edge, "The Sociology of Innovation," p. 331.

70. The lack of appreciation of the details of planetary astronomy and planetary science was cause for some concern for the Working Group on Planetary Science of the most recent Astronomy Survey Committee. Noting that most undergraduates are exposed to planetary research in the context of astronomy survey courses "almost all taught by stellar or extragalactic astronomers," they suggested that "an effort should be made to improve the presentation of this material" (report of the Working Group on Planetary Science, in *Challenges to Astronomy and Astrophysics in the 1980s*, op. cit., p. 108).

71. *Astronomy Transformed*, p. 397.

72. See Paul Forman, "Beyond Quantum Electronics," for the role of the project officer.

73. See John Law, "Technology and Heterogeneous Engineering: The Case of Portuguese Expansion," pp. 111–34 in Bijker, et al., *The Social Construction of Technological Systems*, op. cit.

74. Michel Callon, "Society in the Making: The Study of Technology as a Tool for Sociological Analysis," pp. 83–103 in Bijker, Hughes, and Pinch, eds., op. cit.

75. Hughes, "The Evolution of Large Technological Systems" and John Law, "Technology and Heterogeneous Engineering: The Case of Portuguese Expansion," both in Bijker, Hughes, and Pinch, eds., op. cit.

76. For Latour, according to Steven Shapin in his essay review of Latour's *Science in Action,* rhetoric is the "general-issue weapon of scientific armies." Shapin, "Following Scientists Around," *Social Studies of Science* 18 (1988): 533–50.

77. Hughes, "The Seamless Web," p. 285.

78. Bijker, Hughes, and Pinch, p. 13.

79. This quotation is often used without a specific citation. I have taken the wording from p. v of Arnold Levine, *Managing NASA in the Apollo Era,* op. cit.

Index